FUNCTION OF THE DAY®

FUN INTERACTIVE BELL WORK FOR HIGH SCHOOL MATH CLASSES

FOR ALGEBRA 2 & UP

Debra Richardson

Edited by Jackie Seddon

Front Cover by Kathryn Parsley

© Debra Richardson 2022

About the Author ~ Debra Richardson

I have been teaching math for the past 30 years, ranging from 7th grade to college, in a variety of schools, in Kentucky, Texas, Oklahoma, and Florida. I enjoyed my experience at a small K-12 school as well as a large high school where I taught over 200 students. Each school brought opportunity to teach students with different backgrounds, needs, and learning styles. I frequently attended workshops to learn new ideas to help my students.

In 2016, I attended a training that inspired an idea for bell work, which I call Function of the Day® and it became the 5-minute bell work that I would use every day since, in every math class I taught – Algebra 1, Algebra 2, Advanced Topics Math, Pre-Calculus, Financial Algebra, and College Algebra. Over the last several years, I saw such amazing results with the students in my classroom from using Function of the Day, that I felt the need to share this idea with other math teachers.

When I first began using Function of the Day for bell work, I had my students keep the functions numbered and organized in a composition notebook. Given an equation, my students graphed the function and wrote ten facts about it in their notebook. I noticed when they couldn't remember something, they would look back in their notebook, and it greatly aided their learning. As we moved to different types of functions, my students remembered the concepts and applied them to new functions. Many of the Function of the Day concepts matched the standards my students were learning in class. Their understanding of the material increased, their vocabulary increased, their retention increased, and they remembered concepts from the beginning of the year. My students did exceptionally well on their End of Course state tests that year.

As I added partner and group discussions with Function of the Day, I could see even greater benefits. As students talked with each other, they verbalized new terminology, showing each other and me their understanding of the concepts. In my lower-level classes, including students with learning disabilities or language learning students, I saw success through the continued repetition of the concepts. These students, who previously struggled or lacked confidence in math, thrived with the bell work and volunteered to participate because they were confident with the material. In my honors classes, Function of the Day gave students the opportunity for abstract thinking to find the most facts. It also challenged them to complete facts quickly given a time limit. Function of the Day has worked for ALL my students of ALL levels in ALL the different classes I have taught. In schools/classes using laptops, my students did their Function of the Day bell work digitally using PowerPoint and submitted their work through Google Classroom which was great for teacher grading. Students who struggled physically writing and drawing graphs became a great at Function of the Day because they could create slides, insert a graph from Desmos, and type the facts.

Another great benefit I saw from using Function of the Day was an increase in student ACT and SAT test scores. While preparing for these tests my students noticed how test practice problems contained the same terminology used with Function of the Day. Many of their test scores increased and they accredited that to Function of the Day. Several of my former high school students who took college math one to two years after high school, said they could still remember Function of the Day and it helped them in their college math class.

The beauty of Function of the Day is that it can be modified to meet the needs of your students and your class. The order can be adjusted to fit your curriculum. Teachers can also add extra functions to meet specific needs. I wrote my first book Function of the Day, to follow an Algebra 2 curriculum, but it can be used for most any high school math class. At the request of many teachers, two months later, I wrote another book Function of the Day for Algebra 1, which focuses on linear equations, inequalities, systems of equations, quadratics, and includes systems of inequalities and piecewise functions. I love introducing Function of the Day to math teachers at conferences across the nation to ultimately benefit many students of all levels.

Function of the Day is fun and interactive and has great benefits to student learning that is adaptable for middle & high school math classes. To quote an educator who saw this bell work in action daily, "Function of the Day is the BEST bell work and should be the standard throughout the nation!" Try it out with your students. It is certainly worth a few minutes a day!

Table of Contents

~ Benefits of Function of the Day® ~
~ How to Use Function of the Day® ~
~~ Testimonials ~~
~ Examples of Student Work ~

FOD #1 – 12 ~ ~ ~ ~ ~ ~ ~ ~ ~ ~ ~ ~ ~ ~ ~ Linear Functions

FOD #13 – 16 ~ ~ ~ ~ ~ ~ ~ ~ ~ ~ System of Linear Functions

FOD #17 – 28 ~ ~ ~ ~ ~ ~ ~ ~ ~ ~ ~ ~ Quadratic Functions

FOD #29 – 36 ~ ~ ~ ~ ~ ~ ~ ~ ~ Absolute Value Functions

FOD #37 – 44 ~ ~ ~ ~ ~ ~ ~ ~ ~ ~ Square Root Functions

FOD #45 – 52 ~ ~ ~ ~ ~ ~ ~ ~ ~ ~ ~ ~ ~ Cubic Functions

FOD #53 – 57 ~ ~ ~ ~ ~ ~ ~ ~ ~ ~ ~ ~ Quartic Functions

FOD #58 – 69 ~ ~ ~ ~ ~ ~ ~ ~ ~ ~ ~ ~ ~Rational Functions

FOD #70 – 80 ~ ~ ~ ~ ~ ~ ~ ~ ~ ~ ~ Exponential Functions

FOD #81 – 88 ~ ~ ~ ~ ~ ~ ~ ~ ~ ~ Logarithmic Functions

FOD #89 – 97 ~ ~ ~ ~ ~ ~ ~ ~ ~ Natural Log and e^x Functions

FOD #98 – 104 ~ ~ ~ ~ ~ ~ ~ ~ ~ ~ Cube Root Functions

FOD #105 – 109 ~ ~ ~ ~ ~ ~ ~ ~ ~ ~ ~ Cosine Functions

FOD #106 – 115 ~ ~ ~ ~ ~ ~ ~ ~ ~ ~ ~ ~ Sine Functions

FOD #116 – 120 ~ ~ ~ ~ ~ ~ ~ ~ ~ ~ Tangent Functions

FOD #121 ~ ~ ~ ~ ~ ~ ~ ~ ~ ~ Functions for Valentine's Day

FOD #122 – 124 ~ ~ ~ ~ ~ ~ ~ ~ ~ ~ Functions for Pi Week

FOD #125 ~ ~ ~ ~ ~ ~ ~ ~ ~ Extra Linear System Functions

FOD #126 ~ ~ ~ ~ ~ ~ ~ ~ ~ ~ ~ Extra Quadratic Function

FOD #127 ~ ~ ~ ~ ~ ~ ~ ~ ~ ~ ~ ~ Extra Cubic Function

FOD #128 ~ ~ ~ ~ ~ ~ ~ ~ ~ ~ ~ Extra Quartic Function

FOD #129 – 130 ~ ~ ~ ~ ~ ~ ~ ~ Extra Exponential Functions

~ ~ ~ Appendix of Vocabulary Terms ~ ~ ~

Function of the Day®

is fun interactive daily bell work that WILL...

- **Increase Vocabulary**
  There are over 130 concepts covered with Function of the Day. By understanding and using terminology multiple times, students' vocabulary greatly increases.

- **Increase Math Fluency**
  By building on foundations and using repetition and reasoning, students make connections with function concepts and apply methods for solving problems.

- **Increase Understanding**
  Repetition and verbalization allow students to fully grasp concepts. As students engage in discussions as they analyze facts about functions, their understanding is increased.

- **Increase Retention**
  Retention is gained from frequency. From the repetition of concepts in 130 Functions of the Day, students will retain the information learned and remember it long term.

- **Increase Test Scores**
  With increased understanding of function concepts and vocabulary, students comprehend terminology used on standardized tests and their scores increase. Function of the Day provides a cumulative review that helps fill in gaps and prepares students for the ACT & SAT.

- **Prepare for College Math**
  Concepts learned through Function of the Day are a large part of the curriculum for College Algebra. Understanding these concepts helps students in their college math courses.

Other Benefits

Helps with classroom management. Function of the Day provides routine. Every day as students walk into the classroom, they know the expectation. They get out their notebooks and begin working on the Function of the Day. They draw a graph and write facts as they analyze the function.

Gives opportunity for collaboration that "facilitates meaningful mathematical discourse" (NCTM Standard 4). As students collaborate together using function terminology, they learn from each other and gain a greater understanding of function concepts. Observing students in small group discussions allows the teacher to gauge students' levels of comprehension.

EVERY student can do it! Function of the Day is bell work that everyone can do. Because Function of the Day increases in difficulty as it builds daily on previous concepts, it is excellent for struggling students, ESE and ELL students. Function of the Day is also good for advanced students because it allows for abstract thinking. Because of the routine and repetition of concepts from Function of the Day, students will actively participate because they know what to do and how to do it.

How to Use Function of the Day®

Step 1: Post the equation on the board.

- Include the date, Function of the Day number, and equation on the board. For the first function, post the following: **Date** _____ **FOD #1** $f(x) = $ _____

Step 2: Students draw the graph, analyze the function and write at least 10 facts about the function.

- Facts may include function type, table, slope, intercepts, increasing/decreasing, domain & range, quadrants, parent function, transformations, symmetry, asymptotes, etc.
- Students can be given time independently or in small groups to complete facts.

Step 3: As a class, check facts, discuss new concepts & vocabulary, and correct misunderstandings.

- Teacher can have students volunteer to write facts on the board or assign students to write specific facts on the board.
- Teacher can show answer key for students to check facts.

Recommendations:

- Discuss each parent function together with the class since new concepts are introduced with different function types and explanation of new terms is essential.
- Require students to keep a notebook, numbered and organized with the Functions of the Day and receive a grade for their work. Function of the Day notebook may also be done digitally.
- Use as bell work four days per week to allow one day each week for a quiz or activity.
- Alter the order of functions to fit curriculum. Modify to fit the needs of the students.

Variations (Different ways to do Function of the Day®)

- Students can work independently or with a partner.
- Students can work in small groups (3 or 4) which is great for collaboration and discussion.
- Students can compete in small groups to be the first group to get all ten function facts
- Students can compete in small/large groups to get the most facts within a time limit. Entire classes could compete against other classes. It is possible to get 100 facts for some functions.
- Students can create a poster or anchor chart with facts to display in the classroom.

Student Testimonials

Hello Ms. Richardson!!! I was just looking at my grades from my first mid-report at V. College and I just wanted to say I am so appreciative that I was in your class for the past 2 years. Your class prepared me for college classes more than any other class I have ever taken in high school. You are part of the reason I am passing Pre-Cal, Trig, and Stats, all with 100%s. It's all because of those Function of the Days! I am beyond thankful for you and the impact you have made on my life. You will always be my favorite teacher. Thank you!   – Caleb M.

The function of the day in Mrs. Richardson's class is something we did every day in her class. It became such a routine and has really helped me in my college algebra class that I am taking. The rhymes, songs, and poems still come to mind when I am working on a certain problem. Thank you Mrs. Richardson I will always cherish your class!!   – Carlie B.

Before Mrs. Richardson's class I didn't really understand functions. I like having Mrs. Richardson as a teacher because she doesn't mind repeating things to help us understand. Function of the day emphasizes that repetition is key. And that what's math really is. Not only did function of the day help me in Mrs. Richardson's Advanced Topics class, it also helped me in my Algebra 2 class.   – Renee L.

I was taking College Algebra by dual enrollment during my second semester of high school while in Mrs. Richardson's Advanced Topics class. While doing my unit on quadratics in College Algebra, function of the day helped me remember how to find the vertex, axis of symmetry, and especially with the domain and range. Function of the day is really helping me in my College Algebra class.   – Ramon Z.

I always love FOD. It's a simple way to learn all the topics essential for success in any math course. It has even allowed me to score 100 points higher on my SAT. Thanks so much Mrs. Richardson!   – Daniel G.

Dear Mrs. Richardson, I wanted to say thank you for everything this year. The past few years haven't been the easiest when it has come to math. This year there hasn't been a time where I have felt confused or frustrated in math. You have made learning this year simple and fun. I have always loved math until there was too much going on and it felt as though I wasn't being taught. The function of the day has made such a huge difference for me. Being able to go over and over different things everyday has made everything else be so much easier to understand. Function of the day has given me confidence in math. I wish function of the day was used before this year. I'm so glad I was able to take your class this year!   – Raine F.

I had been in upper-level math courses and then took Mrs. Richardson's class where we did function of the day. Function of the day helped me remember older material that I had not used for several years, and it helped improve my SAT math by 100.   – Alex O.

Hey Mrs. Richardson! I wanted to let you know that I really appreciated your class. In December I did a winter internship and had to fill out a W-4 and knew how to fill it out because of what we practiced in class. I also was very appreciative of the functions of the day. I took college algebra and passed the class with a 97. I would not have been successful in college without taking Advanced Topics Math with our functions of the day. Thank you for all your help and making the best cookies ever!   – Kelsey H.

Function of the day is something that has helped me so much. It has unleashed my understanding of math and it has made me feel as smart as ever because it's so easy. Function of the day has helped me with me with my standardized tests, like the ACT and SAT. There were so many questions where I said, "OH MY GOD, this is literally function of the day." It is something I truly recommend!   – Jose G.

Teacher and Parent Testimonials

As a support teacher in an upper-level math course, Debra Richardson's Function of the Day notebook is an excellent tool to help students strengthen their math skills. Students start the first day with a simple mathematics problem. Through her book, she teaches you how to help the students break down the problem using math terms and sentences. This allows students to identify as many facts as they can about the problem. Each subsequent day builds on the problem to create a complex task they can understand with knowing the foundation. With time, math is no longer a scary word and students build confidence in learning what math problems are asking. I absolutely loved learning and using this tool especially as a non-mathematics teacher.   – Mrs. Neenan (Florida)

As a special education teacher that has taught Algebra I and II, and co-taught in many Algebra I and Algebra II classrooms, I have found the Function of the Day to be an excellent addition to any of these curriculums, and it is especially useful for Specific Learning Disabled students in general education curriculum algebra classrooms. The Function of the Day provides students structure and classroom routine, which is invaluable when teaching SLD students. Students know class begins each day with a Function of the Day, helping them to focus, have materials ready, and prepare for instruction. The Function of the Day reviews a previously taught concept and provides extra reinforcement and practice on this concept. Most importantly, it provides a successful math experience for students that may struggle with algebra concepts, resulting in increased self-confidence.   – Mrs. Culver (Oklahoma)

Function of the Day is a game changer. I used it in my Algebra and Geometry classes. By the beginning of March, I was a month ahead in my Algebra classes. I taught radical functions for the first time in Algebra 1. My students are so comfortable with translations, reflections, stretches, compressions, and all of the vocabulary related to functions. I can give them a new type of function and they can figure most of it out on their own. Since my Geometry students had not done this last year, I also used it with them. It was a great review and helped solidify their knowledge as well.... I am so glad I wandered into your session last summer at the Oklahoma Math Teacher Conference.   – Melissa (Oklahoma)

I met Ms. Richardson at FCTM 2 years ago where she told me about Function of the Day. This past year I used Function of the Day in my classes. Her book provided not only function facts and vocabulary for each day, but various ways that Function of the Day could be used. I was able to choose the way that best fit me, and I adapted it to my students.   – Debbie (Florida)

Mrs. Richardson, I just want to thank you for whatever magic you are working in your math class. My daughter, who is smart and has always loved learning, has unfortunately been left very frustrated these past few years. Since my daughter has had you as a teacher I have seen a phenomenal difference in her attitude towards and confidence in mathematics. Not only is she learning the skills, but also how and when to apply them in the real world. She has told me about your Function of the Day more times than I can count! She has pointed out different instances where she has directly connected what you've taught in your classroom to something meaningful that she is doing outside of school.... And it's not just my child who is benefitting from your methods, it's other students as well. Just a few weeks ago she showed me a text from one of her friends who is now a freshman in college. He was discussing his classes and coursework with her and specifically said that he frequently used so many of the skills he learned from your "Function of the Day" lessons, and how useful those skills are. So, thank you for helping my daughter find the joy in learning again, and for helping to nurture a resurgence in her confidence in math.   – Jean (Florida)

Sample Student Work: Linear ~ Linear Systems ~ Quadratic

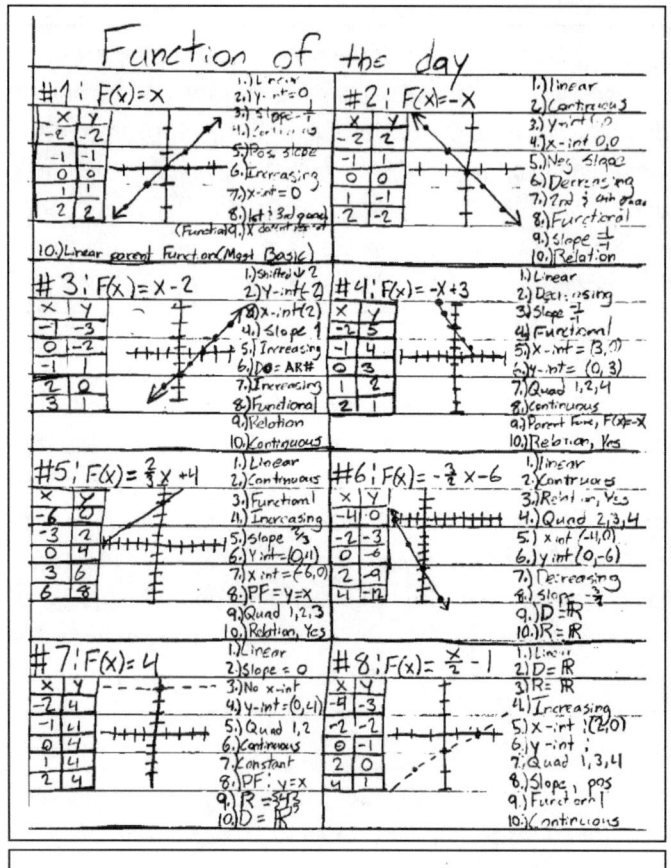

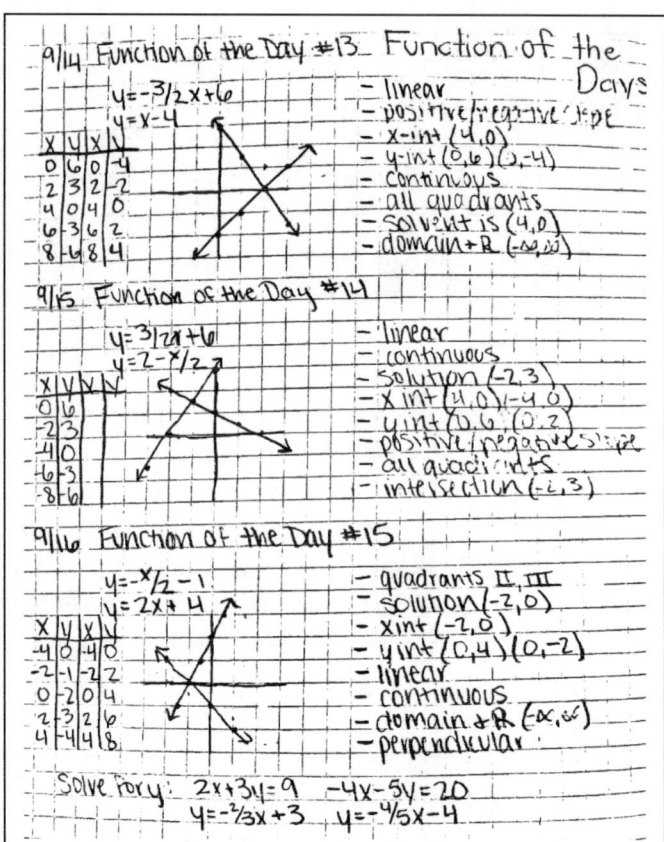

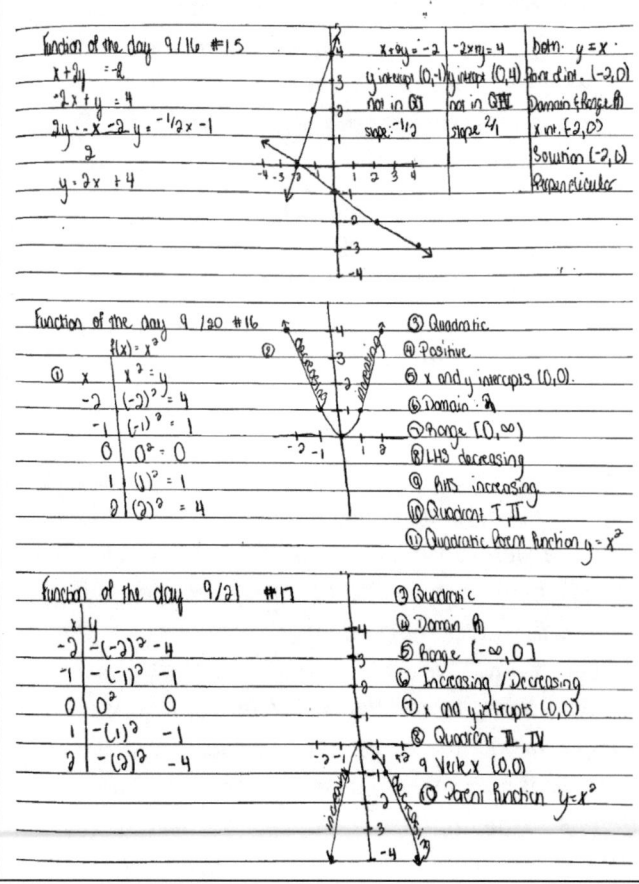

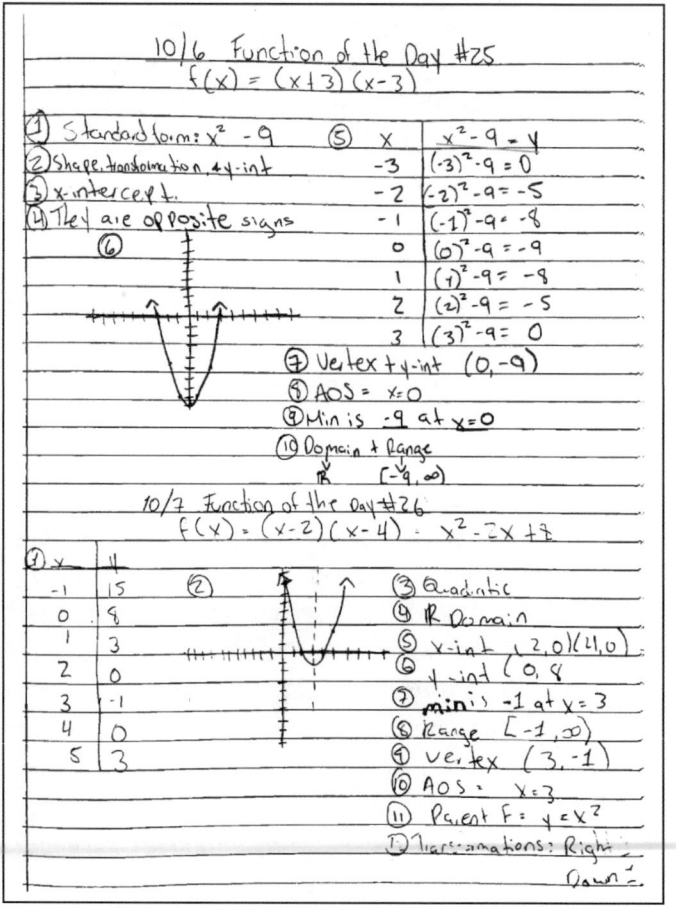

Sample Student Work: Absolute Value ~ Square Root ~ Cubic

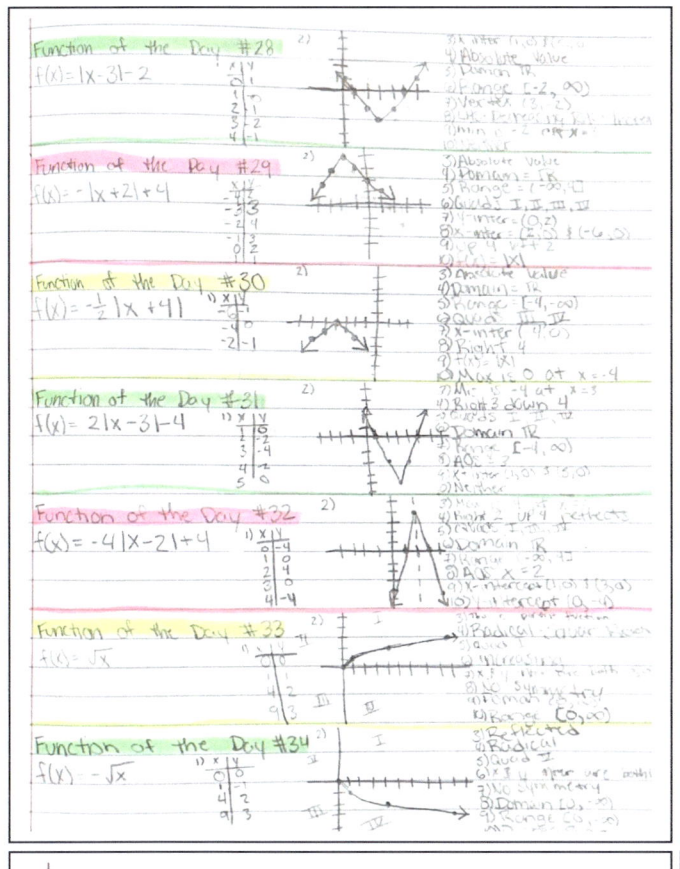

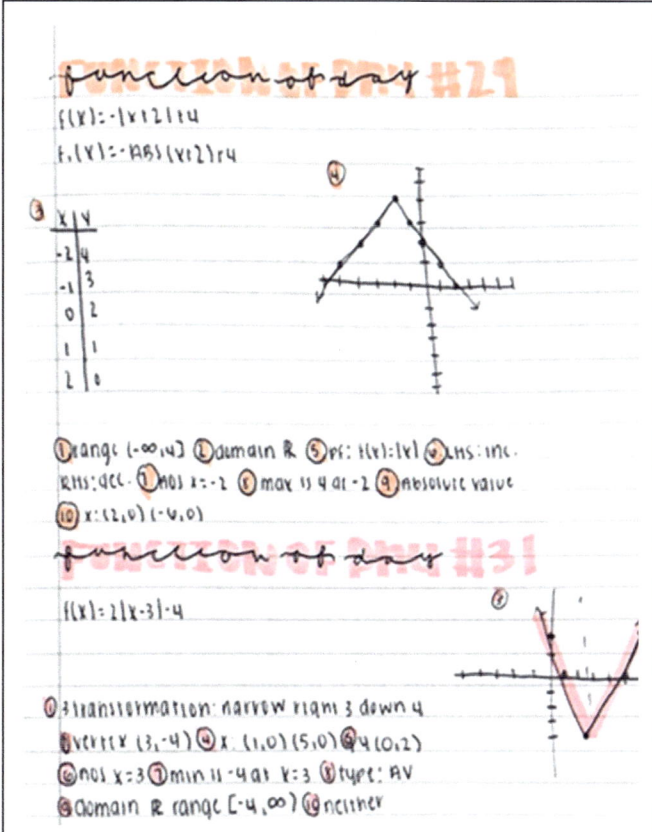

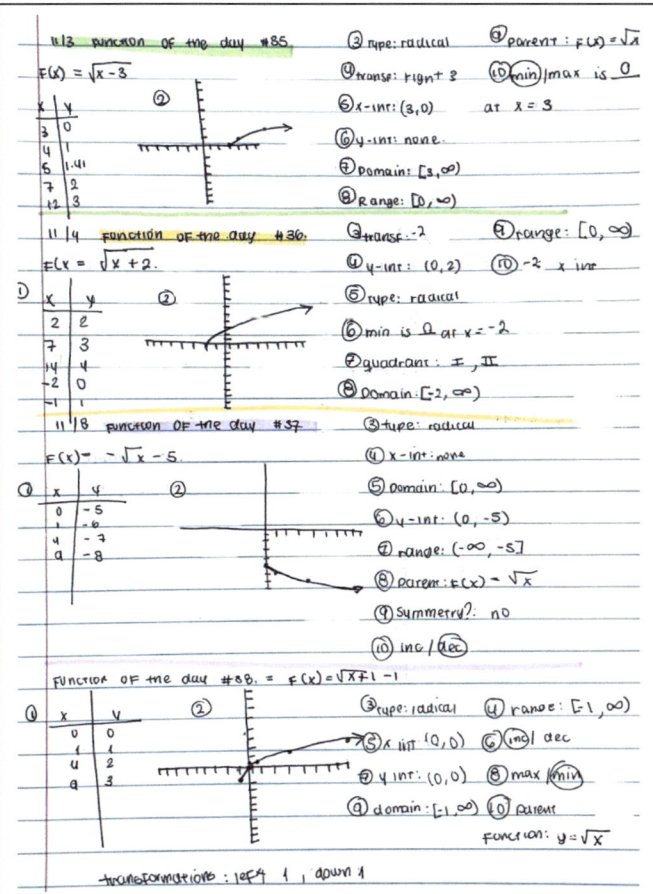

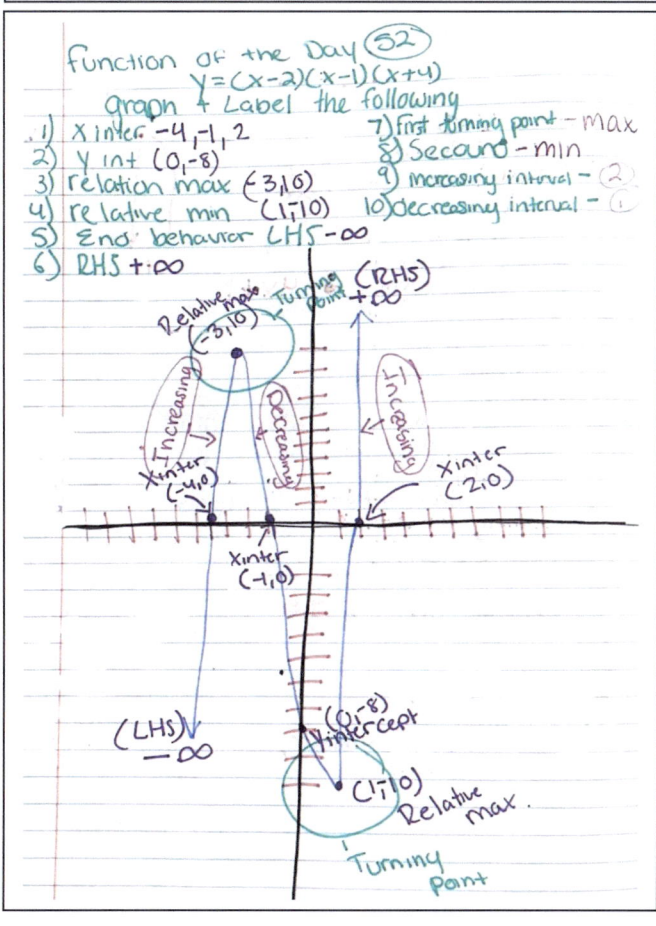

Sample Student Work: Cubic ~ Rational ~ Exponential

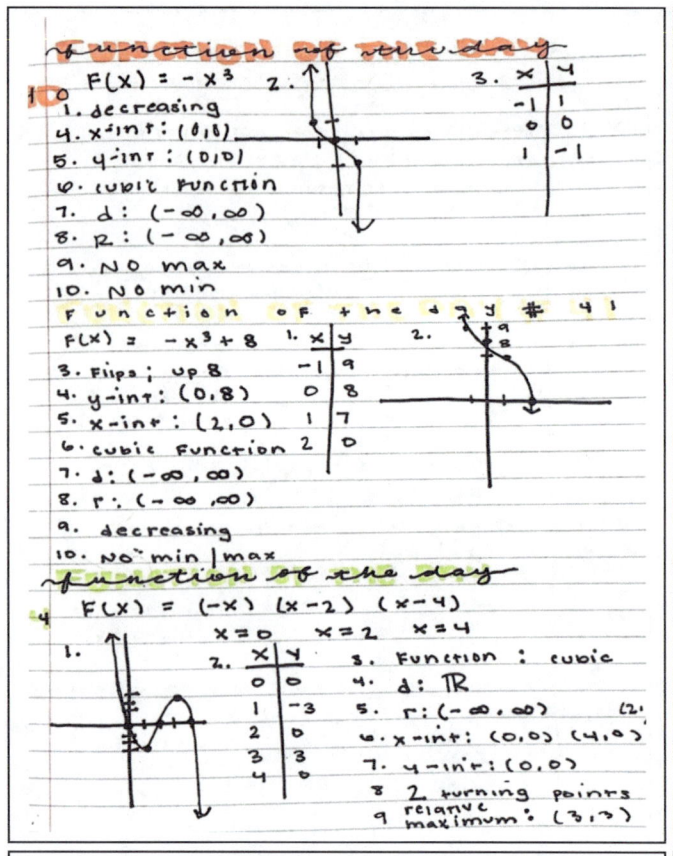

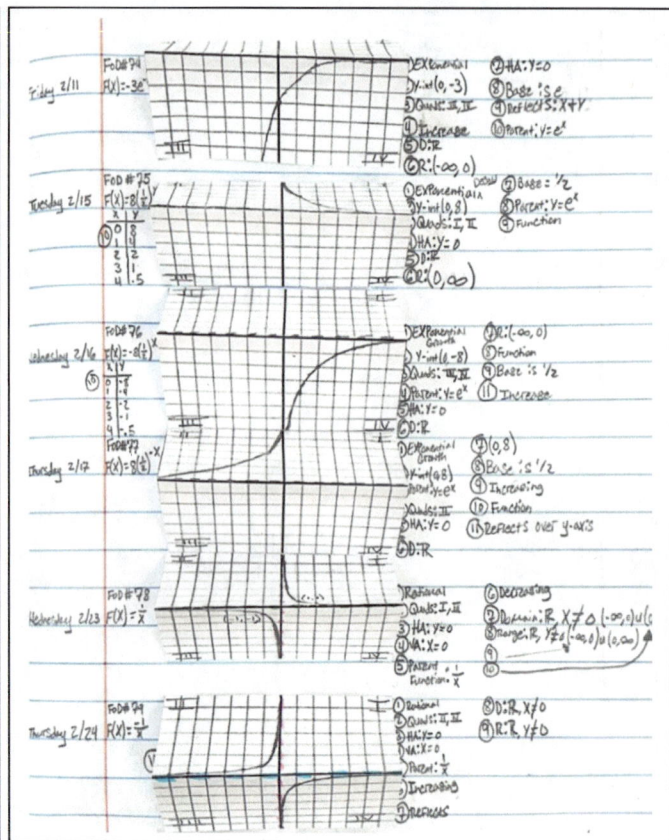

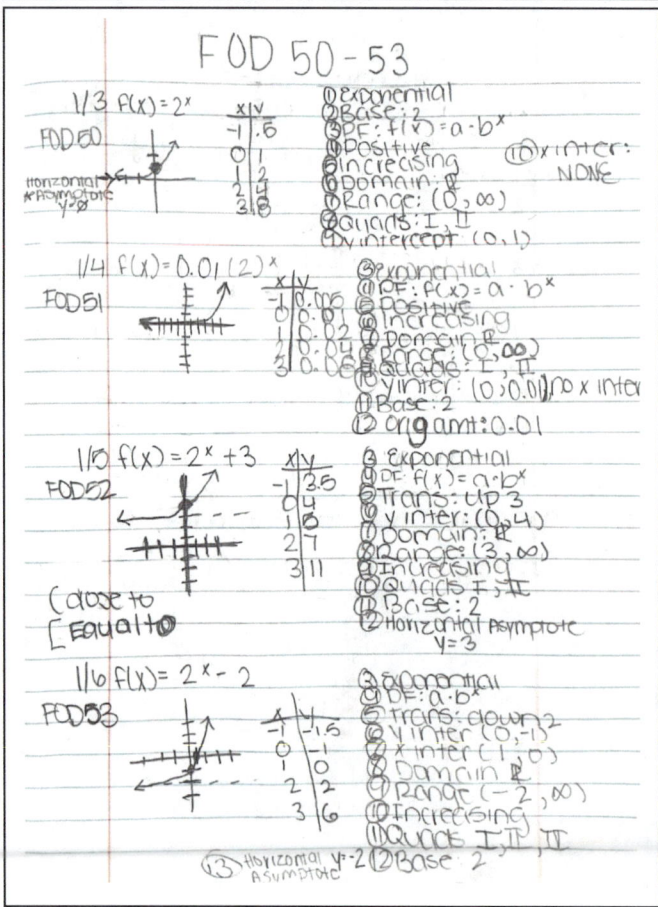

Sample Student Work: Log ~ Natural Log ~ e^x ~ Sine

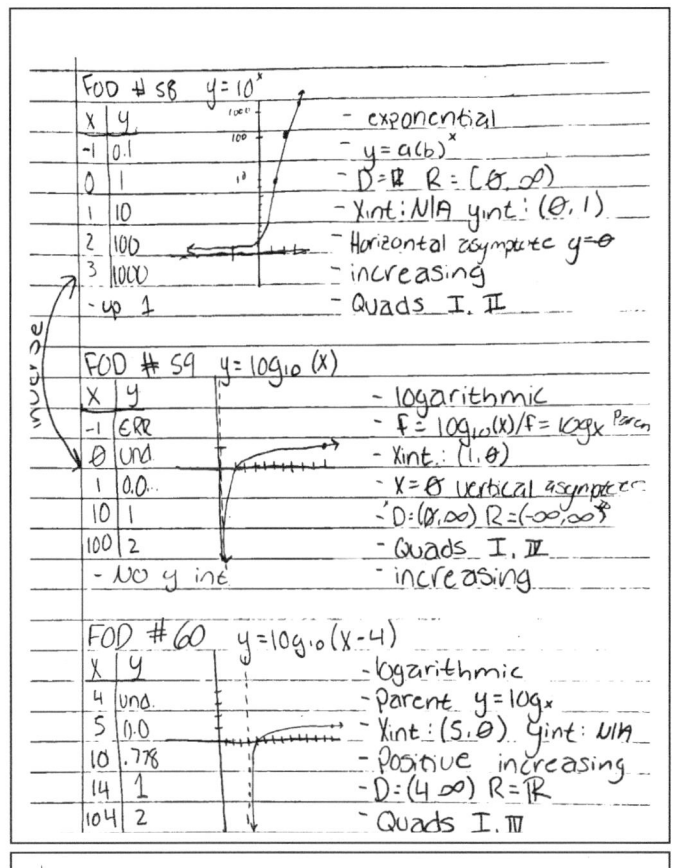

FOD #58 $y = 10^x$

x	y
-1	0.1
0	1
1	10
2	100
3	1000
- up 1

- exponential
- $y = a(b)^x$
- $D = \mathbb{R}$ $R = (0, \infty)$
- Xint: N/A yint: (0, 1)
- Horizontal asymptote $y = 0$
- increasing
- Quads I, II

FOD #59 $y = \log_{10}(x)$

x	y
-1	ERR
0	und
1	0.0
10	1
100	2
- no y int

- logarithmic
- $f = \log_{10}(x) / f = \log x$ Parent
- Xint: (1, 0)
- $x = 0$ vertical asymptote
- $D: (0, \infty)$ $R = (-\infty, \infty)$
- Quads I, IV
- increasing

FOD #60 $y = \log_{10}(x-4)$

x	y
4	und
5	0.0
10	.778
14	1
104	2

- logarithmic
- Parent $y = \log x$
- Xint: (5, 0) yint: N/A
- positive increasing
- $D: (4, \infty)$ $R = \mathbb{R}$
- Quads I, IV

FOD #70
Similarities + Differences
of e^x and $\ln(x)$

HA: $y = 0$ VA: $x = 0$

#70 FOD

Similarities	Differences
Increasing	Range
one nice point	VA / HA
Quad I	LHS → 0, -∞
base of e	Domain
asymptotes = 0	ln x: x-intercept
RHS → +∞	e^x: y-intercept
Graphs reflect y = x	x + y intercepts swapped
Inverses	e^x more rapid increase

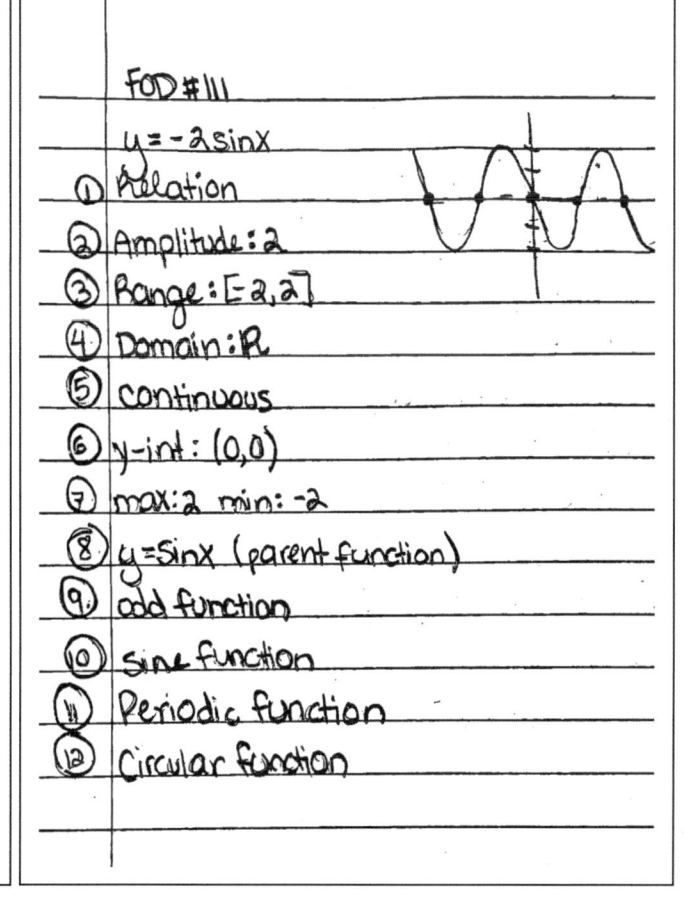

Function of the Day

$y = e^x + 4$ ①   ②

x	y
-10	4
0	5
1	6.7

③ exponential
④ continuous
⑤ domain: $\mathbb{R}$
⑥ range: (4, ∞)
⑦ HA: $y = 4$
⑧ y-int (0, 5)
⑨ no x-int
⑩ quad I, II
⑪ not in quad III, IV
⑫ relation
⑬ function
⑭ no VA
⑮ not ln
⑯ trans. up 4

FOD #111

$y = -2\sin x$

① Relation
② Amplitude: 2
③ Range: [-2, 2]
④ Domain: $\mathbb{R}$
⑤ continuous
⑥ y-int: (0, 0)
⑦ max: 2 min: -2
⑧ $y = \sin x$ (parent function)
⑨ odd function
⑩ sine function
⑪ Periodic function
⑫ Circular function

Functions with Answer Keys

Format for Function of the Day Answer Keys

FOD #_____ $f(x) = $ _____

- Fact 1
- Fact 2
- Fact 3
- Fact 4
- Fact 5
- Fact 6
- Fact 7
- Fact 8
- Fact 9
- Fact 10

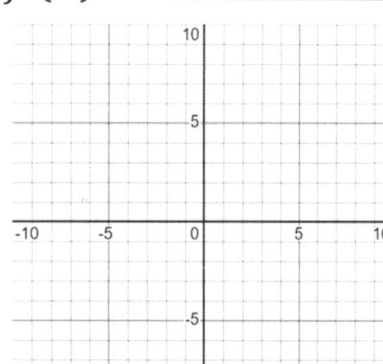

FOD #1 $f(x) = x$

- Linear Function
- Relation
- Continuous
- Positive slope
- Increasing
- Slope is 1
- Crosses through the origin
- y-intercept is (0, 0)
- x-intercept is (0, 0)
- Domain: All Real Numbers $\mathbb{R}$ or (–∞, ∞)
- Range: All Real Numbers $\mathbb{R}$ or (–∞, ∞)
- Quadrants I & III
- Linear Parent Function

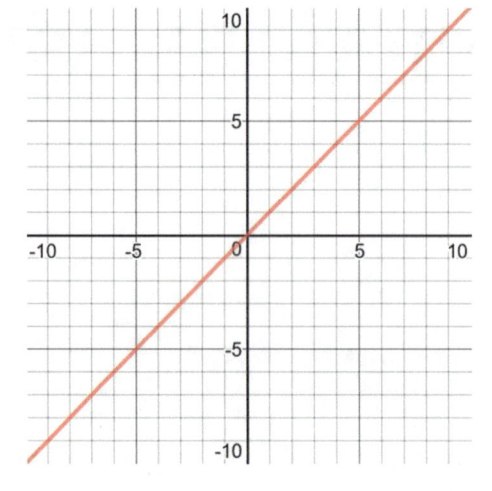

© Debra Richardson 2022

FOD #2 $f(x) = -x$

- Linear Function
- Relation
- Continuous
- Negative Slope
- Decreasing
- Slope is -1
- Crosses through the origin
- y-intercept is (0, 0)
- x-intercept is (0, 0)
- Domain: All Real Numbers $\mathbb{R}$ or (–∞, ∞)
- Range: All Real Numbers $\mathbb{R}$ or (–∞, ∞)
- Quadrants II & IV
- Parent Function is $f(x) = x$

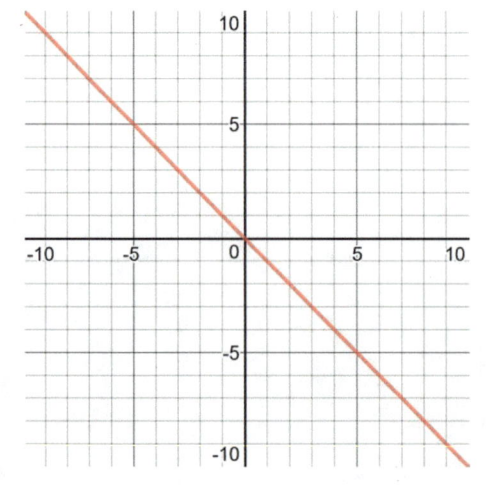

© Debra Richardson 2022

FOD #3

$$f(x) = x - 2$$

- Linear Function
- Relation
- Continuous
- Positive Slope
- Increasing
- Slope is 1
- Crosses both axes
- y-intercept is (0, −2)
- x-intercept is (2, 0)
- Domain: All Real Numbers ℝ or (−∞, ∞)
- Range: All Real Numbers ℝ or (−∞, ∞)
- Quadrants: I, III & IV
- Equation in Slope–Intercept Form
- Parent Function is $f(x) = x$

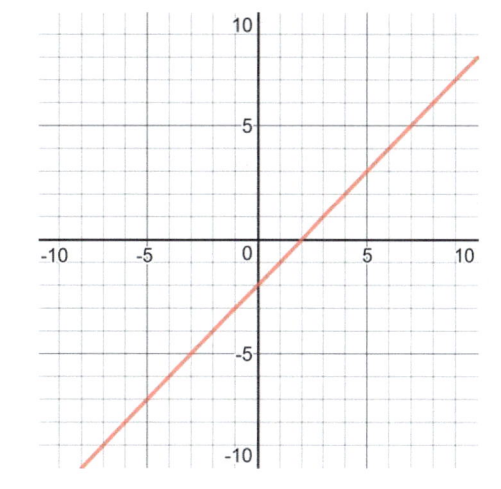

© Debra Richardson 2022

FOD #4

$$f(x) = -x + 3$$

- Linear Function
- Relation
- Continuous
- Negative Slope
- Decreasing
- Slope is −1
- Crosses both axes
- y-intercept is (0, 3)
- x-intercept is (3, 0)
- Domain: All Real Numbers ℝ or (−∞, ∞)
- Range: All Real Numbers ℝ or (−∞, ∞)
- Quadrants I, II & IV
- Equation in Slope–Intercept Form
- Parent Function is $f(x) = x$

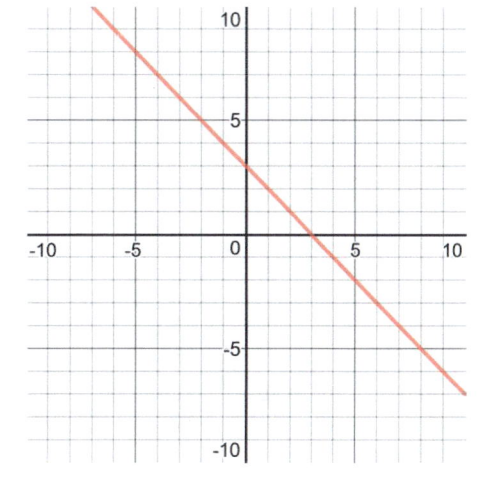

© Debra Richardson 2022

FOD #5

$$f(x) = \frac{2}{3}x + 4$$

- Linear Function
- Relation
- Continuous
- Positive Slope
- Increasing
- Slope is 2/3
- Crosses both axes
- y-intercept is (0, 4)
- x-intercept is (−6, 0)
- Domain: All Real Numbers $\mathbb{R}$ or (−∞, ∞)
- Range: All Real Numbers $\mathbb{R}$ or (−∞, ∞)
- Quadrants I, II & III
- Equation in Slope–Intercept Form
- Parent Function is $f(x) = x$

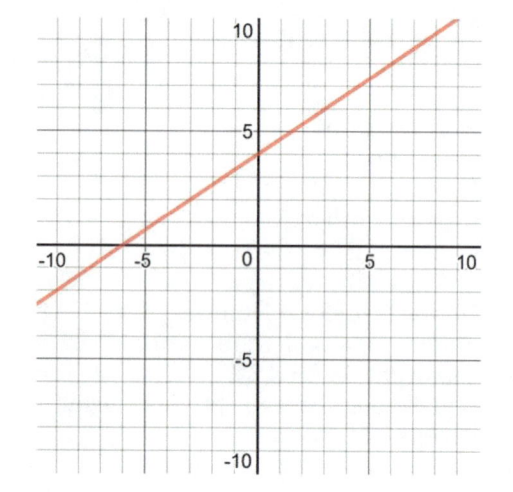

© Debra Richardson 2022

FOD #6

$$f(x) = -\frac{3}{2}x - 6$$

- Linear Function
- Relation
- Continuous
- Negative Slope
- Decreasing
- Slope is −3/2
- Crosses both axes
- y-intercept is (0, −6)
- x-intercept is (−4, 0)
- Domain: All Real Numbers $\mathbb{R}$ or (−∞, ∞)
- Range: All Real Numbers $\mathbb{R}$ or (−∞, ∞)
- Quadrants II, III, & IV
- Equation in Slope–Intercept Form
- Parent Function is $f(x) = x$

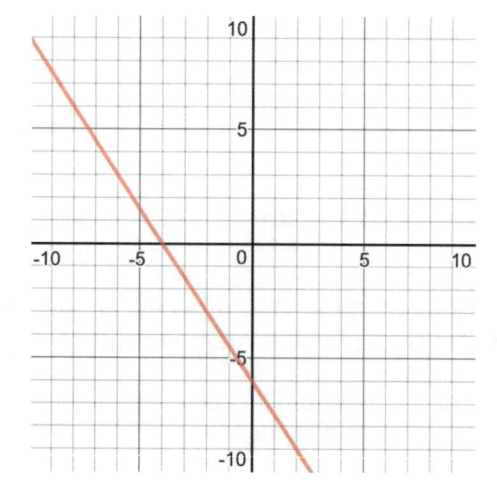

© Debra Richardson 2022

FOD #7

$$f(x) = \frac{x}{2} - 1$$

- Linear Function
- Relation
- Continuous
- Positive Slope
- Increasing
- Slope is ½
- Crosses both axes
- y-intercept is (0, –1)
- x-intercept is (2, 0)
- Domain: All Real Numbers $\mathbb{R}$ or (–∞, ∞)
- Range: All Real Numbers $\mathbb{R}$ or (–∞, ∞)
- Quadrants I, III & IV
- Equation in Slope–Intercept Form
- Parent Function is $f(x) = x$

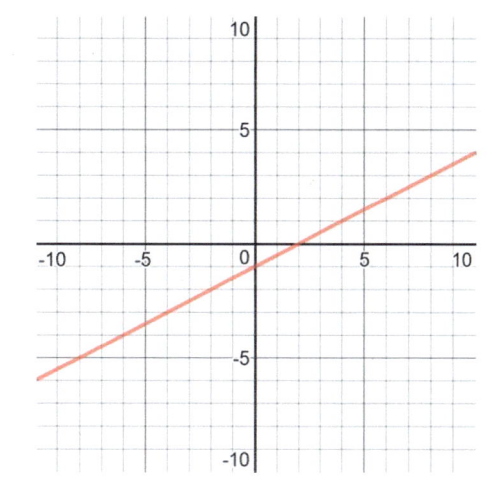

FOD #8

$$f(x) = -4$$

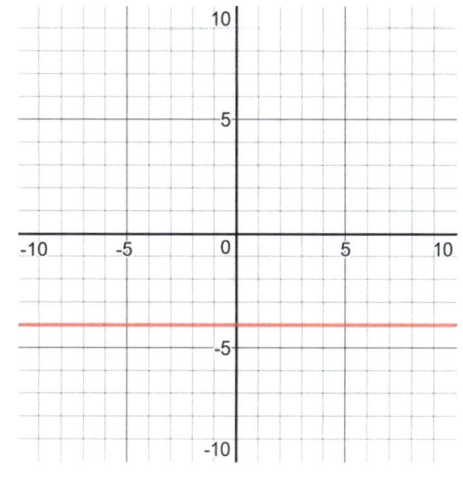

- Linear Function
- Relation
- Continuous
- Horizontal Line
- Constant
- Slope is 0
- Crosses the y-axis
- y-intercept is (0, –4)
- No x-intercept
- Domain: All Real Numbers $\mathbb{R}$ or (–∞, ∞)
- Range: { –4 } or $y = -4$
- Quadrants III & IV
- Not in Quadrants I & II

FOD #9

$$2x + 4y = 8$$

- Linear Function
- Equation in Standard Form
- Slope-Intercept Form is $y = -\frac{1}{2}x + 2$
- Relation, Continuous
- Decreasing
- Slope is –1/2
- Crosses both axes
- y-intercept is (0, 2)
- x-intercept is (4, 0)
- Domain: All Real Numbers $\mathbb{R}$ or (–∞, ∞)
- Range: All Real Numbers $\mathbb{R}$ or (–∞, ∞)
- Quadrants I, II & IV
- Not in Quadrant III
- Parent Function is $f(x) = x$

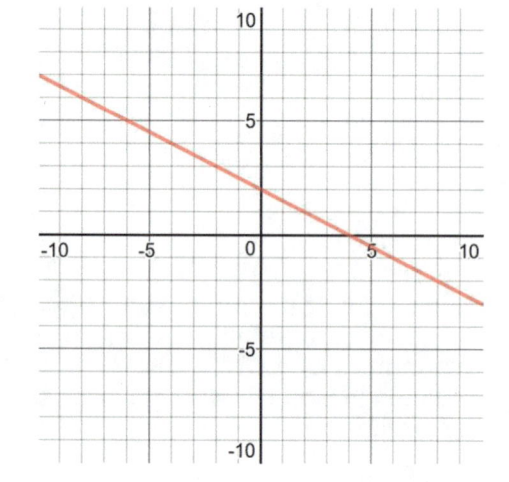

© Debra Richardson 2022

FOD #10

$$x - y = 6$$

- Linear Function
- Equation in Standard Form
- Slope-Intercept Form is $y = x - 6$
- Relation, Continuous
- Increasing
- Slope is 1
- Crosses both axes
- y-intercept is (0, –6)
- x-intercept is (6, 0)
- Domain: All Real Numbers $\mathbb{R}$ or (–∞, ∞)
- Range: All Real Numbers $\mathbb{R}$ or (–∞, ∞)
- Quadrants I, III & IV
- Not in Quadrant II
- Parent Function is $f(x) = x$

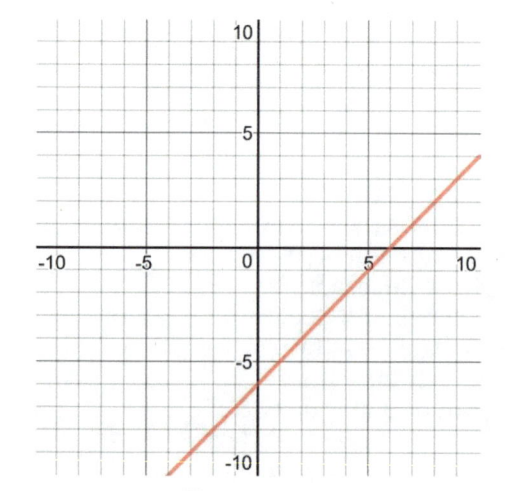

© Debra Richardson 2022

FOD #11 $\qquad -3x - 6y = 12$

- Linear Function
- Equation in Standard Form
- Slope-Intercept Form is $y = -\frac{1}{2}x - 2$
- Relation, Continuous
- Decreasing
- Slope is –1/2
- Crosses both axes
- y-intercept is (0, –2)
- x-intercept is (–4, 0)
- Domain: All Real Numbers $\mathbb{R}$ or (–∞, ∞)
- Range: All Real Numbers $\mathbb{R}$ or (–∞, ∞)
- Quadrants II, III & IV
- Not in Quadrant I
- Parent Function is $f(x) = x$

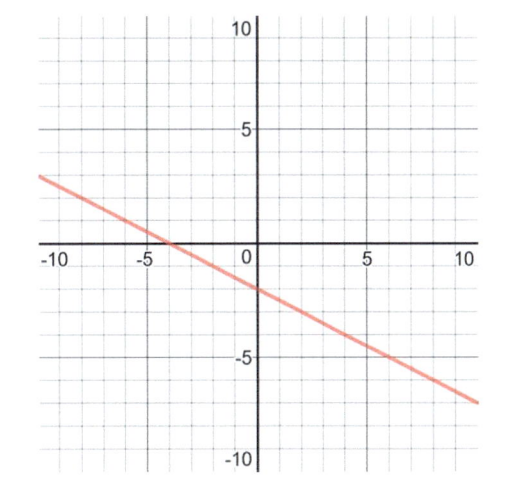

FOD #12 $\qquad 2x - 5y = 10$

- Linear Function
- Equation in Standard Form
- Slope-Intercept Form is $y = \frac{2}{5}x - 2$
- Relation, Continuous
- Increasing
- Slope is 2/5
- Crosses both axes
- y-intercept is (0, –2)
- x-intercept is (5, 0)
- Domain: All Real Numbers $\mathbb{R}$ or (–∞, ∞)
- Range: All Real Numbers $\mathbb{R}$ or (–∞, ∞)
- Quadrants I, III & IV
- Not in Quadrant II
- Parent Function is $f(x) = x$

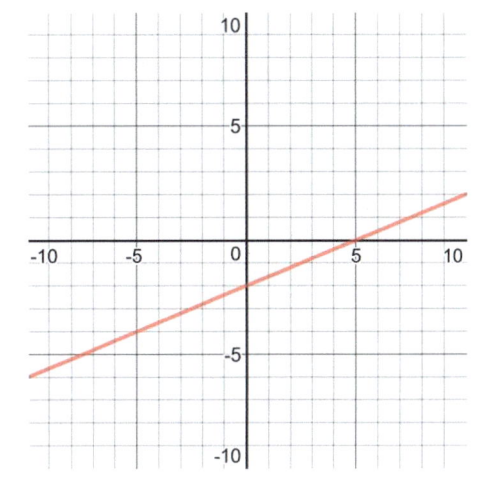

FOD #13

$$y = x + 3$$
$$y = -x + 3$$

- Two Linear Functions
- System of Linear Equations
- Relations
- Continuous
- Lines intersect at (0, 3)
- Solution to the system is (0, 3)
- Increasing and Decreasing Lines
- Slopes are 1 and –1
- Slopes are opposite reciprocals
- Lines are perpendicular
- y-intercept is (0, 3)
- x-intercepts are (–3, 0) and (3, 0)
- Each Domain: All Real Numbers $\mathbb{R}$ or (–∞, ∞)
- Each Range: All Real Numbers $\mathbb{R}$ or (–∞, ∞)

© Debra Richardson 2022

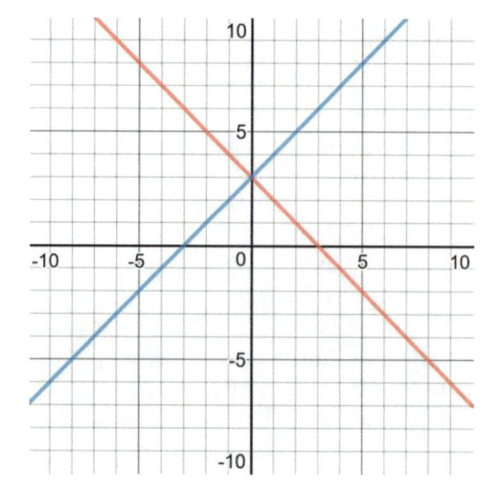

FOD #14

$$y = -\frac{3}{2}x + 6$$
$$y = x - 4$$

- Two Linear Functions
- System of Linear Equations
- Relations
- Continuous
- Lines intersect at (4, 0)
- Solution to the system is (4, 0)
- Decreasing and Increasing Lines
- Slopes are –3/2 and 1
- y-intercepts are (0, 6) and (0, –4)
- Both x-intercepts are (4, 0)
- Each Domain: All Real Numbers $\mathbb{R}$ or (–∞, ∞)
- Each Range: All Real Numbers $\mathbb{R}$ or (–∞, ∞)

© Debra Richardson 2022

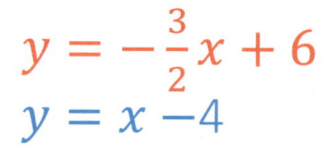

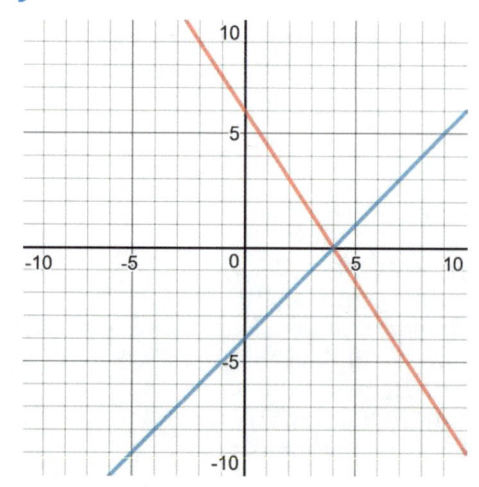

FOD #15

$$y = \frac{3}{4}x + 6$$
$$4y - 3x = -12$$

- Two Linear Functions
- System of Linear Equations
- Slope-Intercept Form of 2nd equation is $y = \frac{3}{4}x - 3$
- Relations
- Continuous
- Lines are parallel
- Lines do not intersect
- No solution to the system
- Increasing Lines
- Both slopes are 3/4
- y-intercepts are (0, 6) and (0, –3)
- x-intercepts are (–8, 0) and (4, 0)
- Each Domain: All Real Numbers $\mathbb{R}$ or (–∞, ∞)
- Each Range: All Real Numbers $\mathbb{R}$ or (–∞, ∞)

© Debra Richardson 2022

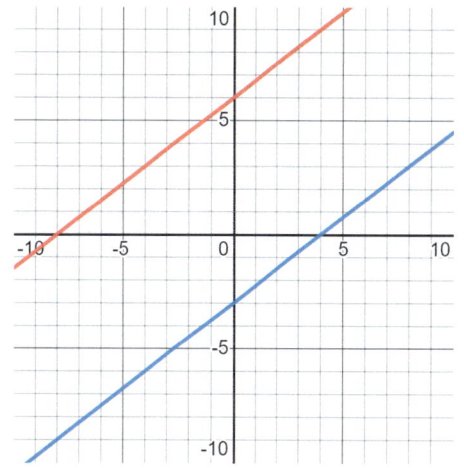

FOD #16

$$x + 2y = -2$$
$$-2x + y = 4$$

- Two Linear Functions
- System of Linear Equations
- Equations are in standard form
- Slope-Intercept Form of equations are $y = -\frac{1}{2}x - 1$ and $y = 2x + 4$
- Relations, Continuous
- Lines intersect at (–2 , 0)
- Solution to the system is (–2, 0)
- Slopes are –1/2 and 2
- Slopes are opposite reciprocals
- Lines are perpendicular
- y-intercepts are (0, –1) and (0, 4)
- Both x-intercepts are (–2 , 0)
- Each Domain: All Real Numbers $\mathbb{R}$ or (–∞, ∞)
- Each Range: All Real Numbers $\mathbb{R}$ or (–∞, ∞)

© Debra Richardson 2022

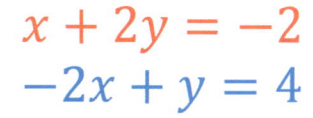

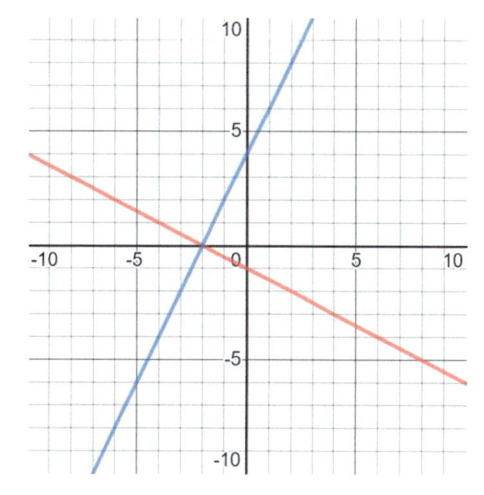

FOD #17

$$f(x) = x^2$$

- Quadratic Parent Function
- Nonlinear, Relation, Continuous
- Parabola opens upward
- Touches the origin
- y-intercept is (0, 0)
- x-intercept is (0, 0)
- Symmetry over y-axis, Even function
- Axis of Symmetry is x = 0
- Vertex is (0, 0)
- Minimum is 0 at x = 0
- Domain: All Real Numbers $\mathbb{R}$ or (–∞, ∞)
- Range: y ≥ 0 or [0,∞)
- Quadrants I & II
- Decreasing Interval (–∞, 0)
- Increasing Interval (0, ∞)

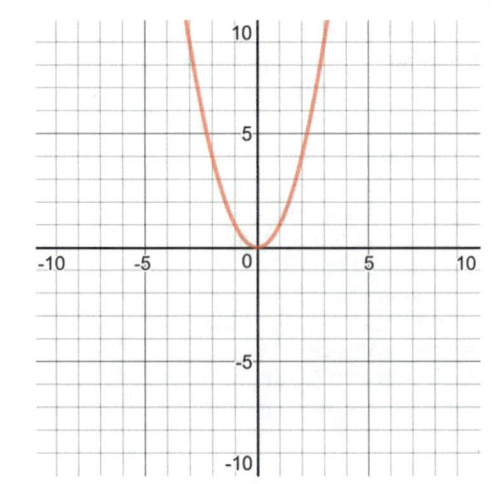

© Debra Richardson 2022

FOD #18

$$f(x) = -x^2$$

- Quadratic Function
- Reflects over x-axis
- Parabola opens downward
- y-intercept is (0, 0)
- x-intercept is (0, 0)
- Symmetry over y-axis, Even function
- Axis of Symmetry is x = 0
- Vertex is (0, 0)
- Maximum is 0 at x = 0
- Domain: All Real Numbers $\mathbb{R}$ or (–∞, ∞)
- Range: y ≤ 0 or (–∞, 0]
- Quadrants III & IV
- Increasing Interval (–∞, 0)
- Decreasing Interval (0, ∞)
- Parent Function $f(x) = x^2$

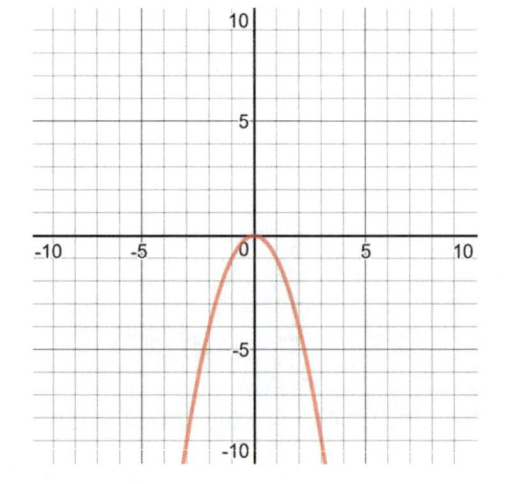

© Debra Richardson 2022

FOD #19

$$f(x) = x^2 + 4$$

- Quadratic Function
- Parabola opens upward
- Vertical shift up 4
- y-intercept is (0, 4)
- No x-intercepts
- Symmetry over y-axis, Even function
- Axis of Symmetry is x = 0
- Vertex is (0, 4)
- Minimum is 4 at x = 0
- Domain: All Real Numbers $\mathbb{R}$ or (–∞, ∞)
- Range: y ≥ 4 or [4, ∞)
- Quadrants I & II; Not in III & IV
- Decreasing Interval (–∞, 0)
- Increasing Interval (0, ∞)
- Parent Function $f(x) = x^2$

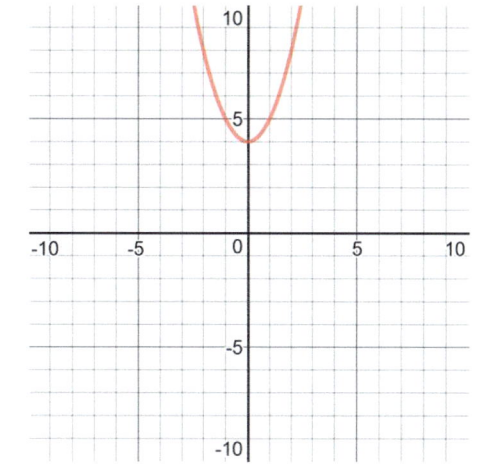

FOD #20

$$f(x) = (x + 5)^2$$

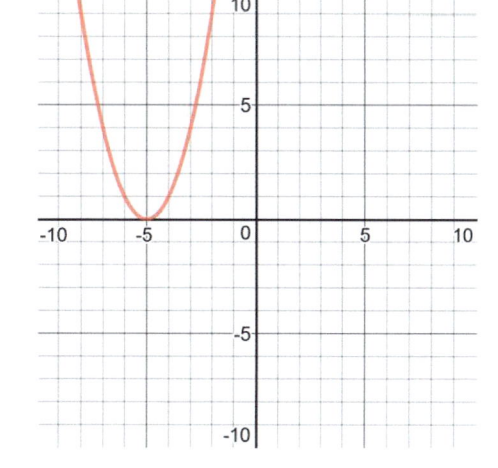

- Quadratic Function
- Standard Form is $f(x) = x^2 + 10x + 25$
- Horizontal shift left 5
- x-intercept is (–5, 0)
- (–5, 0) is a double root
- y-intercept is (0, 25)
- Axis of Symmetry is x = –5
- Vertex is (–5, 0)
- Minimum is 0 at x = –5
- Domain: All Real Numbers $\mathbb{R}$ or (–∞, ∞)
- Range: y ≥ 0 or [0, ∞)
- Quadrants I & II; Not in III & IV
- Decreasing Interval (–∞, –5)
- Increasing Interval (–5, ∞)
- Parent Function $f(x) = x^2$

FOD #21

$$f(x) = (x+2)^2 - 4$$

- Quadratic Function
- Standard Form is $f(x) = x^2 + 4x$
- Horizontal shift left 2
- Vertical shift down 4
- y-intercept is (0, 0)
- x-intercepts are (−4, 0) and (0, 0)
- Axis of Symmetry is x = −2
- Vertex is (−2, −4)
- Minimum is −4 at x = −2
- Domain: All Real Numbers $\mathbb{R}$ or (−∞, ∞)
- Range: y ≥ −4 or [−4, ∞)
- Quadrants I & II, III; Not in IV
- Decreasing Interval (−∞, −2)
- Increasing Interval (−2, ∞)
- Parent Function $f(x) = x^2$

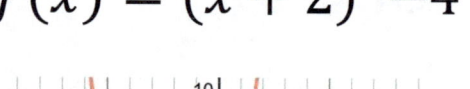

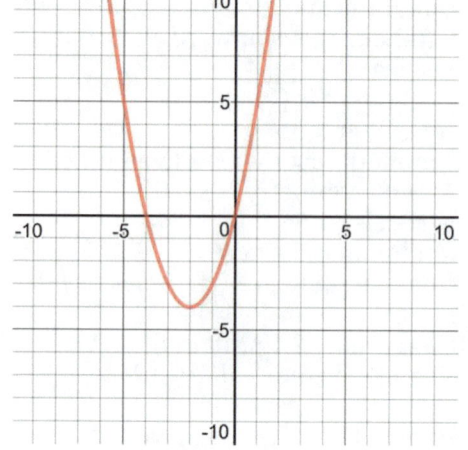

© Debra Richardson 2022

FOD #22

$$f(x) = -(x+1)^2 + 4$$

- Quadratic Function
- Standard Form is $f(x) = -x^2 - 2x + 3$
- Reflects over x-axis
- Horizontal shift left 1
- Vertical shift up 4
- y-intercept is (0, 3)
- x-intercepts are (−3, 0) and (1, 0)
- Axis of Symmetry is x = −1
- Vertex is (−1, 4)
- Maximum is 4 at x = −1
- Domain: All Real Numbers $\mathbb{R}$ or (−∞, ∞)
- Range: y ≤ 4 or (−∞, 4]
- All Quadrants
- Increasing Interval (−∞, −1)
- Decreasing Interval (−1, ∞)
- Parent Function $f(x) = x^2$

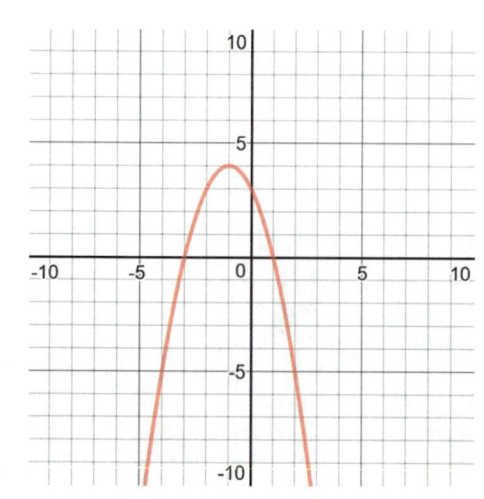

© Debra Richardson 2022

FOD #23 $\qquad f(x) = -5(x-1)^2 + 5$

- Quadratic Function
- Standard Form is $f(x) = -5x^2 + 10x$
- Reflects over x-axis
- Vertical stretch (narrow) by a factor of 5
- Horizontal shift right 1
- Vertical shift up 5
- y-intercept is (0, 0)
- x-intercepts are (0, 0) and (2, 0)
- Axis of Symmetry is x = 1
- Vertex is (1, 5)
- Maximum is 5 at x = 1
- Domain: All Real Numbers ℝ or (−∞, ∞)
- Range: y ≤ 5 or (−∞, 5]
- Quadrants I, III & IV
- Increasing Interval (−∞, 1)
- Decreasing Interval (1, ∞)
- Parent Function $f(x) = x^2$

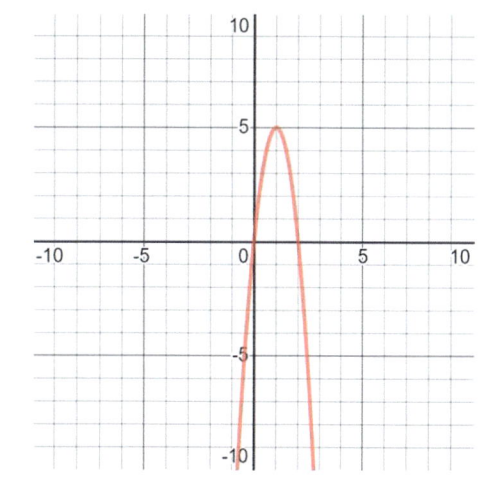

© Debra Richardson 2022

FOD #24 $\qquad f(x) = \frac{1}{2}x^2 - 8$

- Quadratic Function
- Vertical compression (wide) by a factor of ½
- No horizontal shift
- Vertical shift down 8
- y-intercept is (0, −8)
- x-intercepts are (−4, 0) and (4, 0)
- Axis of Symmetry is x = 0
- Vertex is (0, −8)
- Minimum is −8 at x = 0
- Domain: All Real Numbers ℝ or (−∞, ∞)
- Range: y ≥ −8 or [−8, ∞)
- All Quadrants
- Decreasing Interval (−∞, 0)
- Increasing Interval (0, ∞)
- Parent Function $f(x) = x^2$

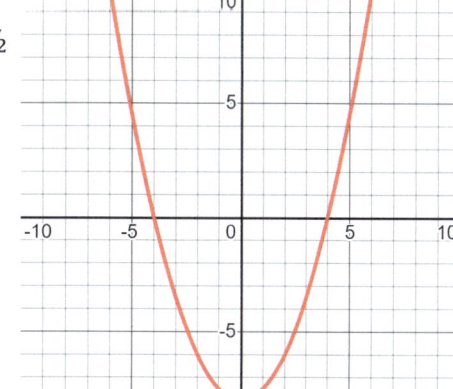

© Debra Richardson 2022

FOD #25

$$f(x) = -x(x-8)$$

- Quadratic Function
- Standard Form is $f(x) = -x^2 + 8x$
- Reflects over x-axis
- y-intercept is (0, 0)
- x-intercepts are (0, 0) and (8, 0)
- Axis of Symmetry is x = 4
- Vertex is (4, 16)
- Maximum is 16 at x = 4
- Horizontal shift right 4
- Vertical shift up 16
- Domain: All Real Numbers $\mathbb{R}$ or (−∞, ∞)
- Range: y ≤ 16 or (−∞, 16]
- Quadrants I, III & IV
- Increasing Interval (−∞, 4)
- Decreasing Interval (4, ∞)
- Parent Function $f(x) = x^2$

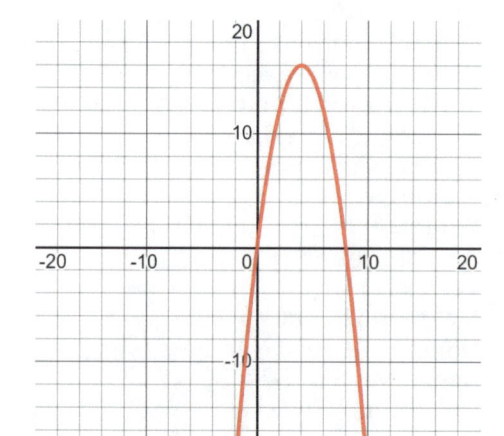

© Debra Richardson 2022

FOD #26

$$f(x) = (x+3)(x-3)$$

- Quadratic Function
- Standard Form is $f(x) = x^2 - 9$
- Parabola opens upward
- Vertical shift down 9
- y-intercept is (0, −9)
- x-intercepts are (−3, 0) and (3, 0)
- Symmetry over y-axis
- Axis of Symmetry is x = 0
- Vertex is (0, −9)
- Minimum is −9 at x = 0
- Domain: All Real Numbers $\mathbb{R}$ or (−∞, ∞)
- Range: y ≥ −9 or [−9, ∞)
- All Quadrants
- Decreasing Interval (−∞, 0)
- Increasing Interval (0, ∞)
- Parent Function $f(x) = x^2$

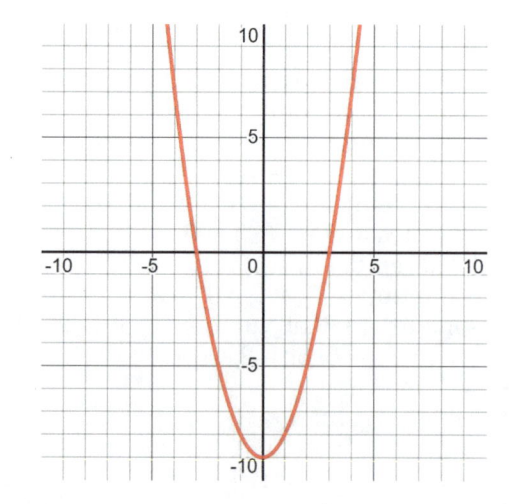

© Debra Richardson 2022

FOD #27

$$f(x) = x^2 + 8x + 16$$

- Quadratic Function
- Factored Form is $f(x) = (x + 4)^2$
- Parabola opens upward
- Horizontal shift left 4
- y-intercept is (0, 16)
- x-intercept is (−4, 0)
- (−4, 0) is a double root
- Vertex is (−4, 0)
- Axis of Symmetry is x = −4
- Minimum is 0 at x = −4
- Domain: All Real Numbers $\mathbb{R}$ or (−∞, ∞)
- Range: y ≥ 0 or [0, ∞)
- Quadrants I & II
- Decreasing Interval (−∞, −4)
- Increasing Interval (−4, ∞)
- Parent Function $f(x) = x^2$

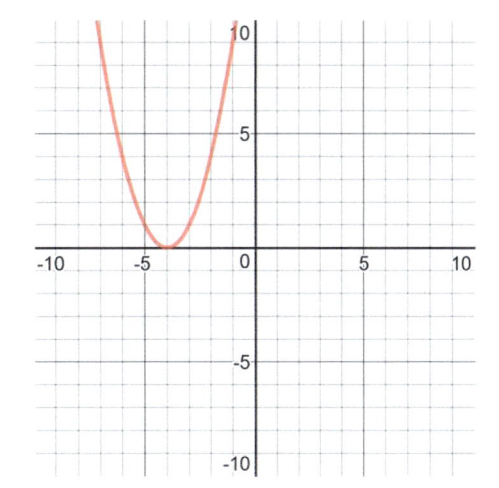

© Debra Richardson 2022

FOD #28

$$f(x) = x^2 - 6x + 8$$

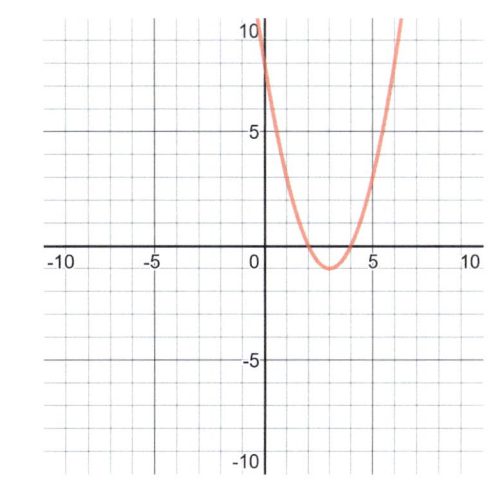

- Quadratic Function
- Factored Form is $f(x) = (x - 2)(x - 4)$
- Vertex Form is $f(x) = (x - 3)^2 - 1$
- y-intercept is (0, 8)
- x-intercepts are (2, 0) and (4, 0)
- Axis of Symmetry is x = 3
- Vertex is (3, −1)
- Minimum is −1 at x = 3
- Horizontal shift right 3
- Vertical shift down 1
- Domain: All Real Numbers $\mathbb{R}$ or (−∞, ∞)
- Range: y ≥ −1 or [−1, ∞)
- Quadrants I, II & IV
- Decreasing Interval (−∞, 3)
- Increasing Interval (3, ∞)
- Parent Function is $f(x) = x^2$

© Debra Richardson 2022

FOD #29

$$f(x) = |x|$$

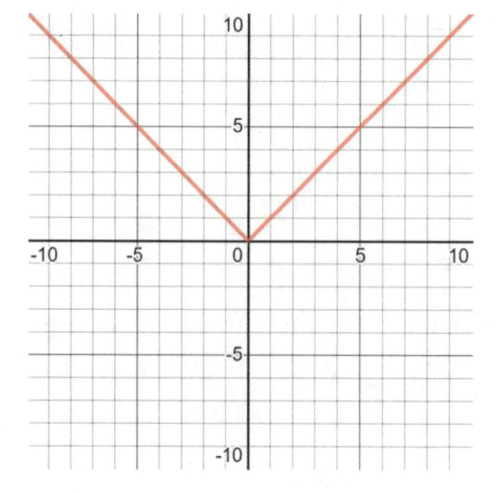

- Absolute Value Parent Function
- Touches the origin
- y-intercept is (0, 0)
- x-intercept is (0, 0)
- Symmetry over y-axis
- Axis of Symmetry is x = 0
- Vertex is (0, 0)
- Minimum is 0 at x = 0
- Domain: All Real Numbers $\mathbb{R}$ or (–∞, ∞)
- Range: y ≥ 0 or [0, ∞)
- Quadrants I & II
- Decreasing Interval (–∞, 0)
- Increasing Interval (0, ∞)

© Debra Richardson 2022

FOD #30

$$f(x) = |x - 3| - 2$$

- Absolute Value Function
- Horizontal shift right 3
- Vertical shift down 2
- Vertex is (3, –2)
- y-intercept is (0, 1)
- x-intercepts are (1, 0) and (5, 0)
- Axis of Symmetry is x = 3
- Minimum is –2 at x = 3
- Domain: All Real Numbers $\mathbb{R}$ or (–∞, ∞)
- Range: y ≥ –2 or [–2, ∞)
- Quadrants I, II & IV
- Decreasing Interval (–∞, 3)
- Increasing Interval (3, ∞)
- Parent Function is $f(x) = |x|$

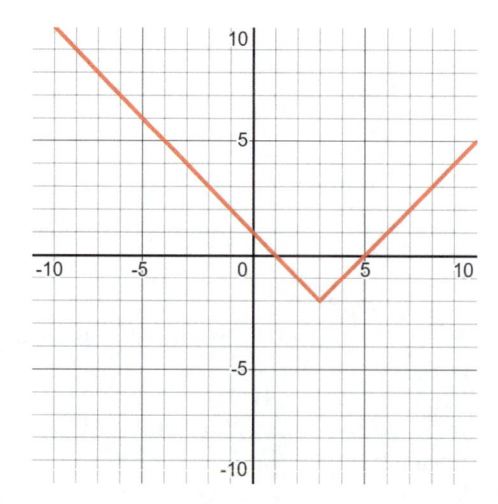

© Debra Richardson 2022

FOD #31

$$f(x) = |x + 4| - 3$$

- Absolute Value Function
- Horizontal shift left 4
- Vertical shift down 3
- Vertex is (−4, −3)
- y-intercept is (0, 1)
- x-intercepts are (−7, 0) and (−1, 0)
- Axis of Symmetry is x = −4
- Minimum is −3 at x = −4
- Domain: All Real Numbers ℝ or (−∞, ∞)
- Range: y ≥ −3 or [−3, ∞)
- Quadrants I, II & III
- Decreasing Interval (−∞, −4)
- Increasing Interval (−4, ∞)
- Parent Function is $f(x) = |x|$

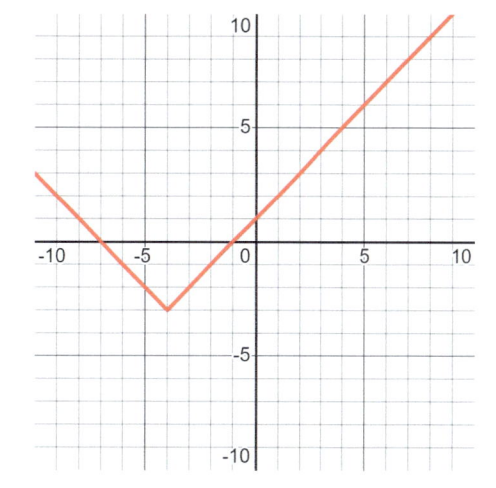

FOD #32

$$f(x) = -|x + 2| + 4$$

- Absolute Value Function
- Reflects over x-axis
- Horizontal shift left 2
- Vertical shift up 4
- Vertex is (−2, 4)
- y-intercept is (0, 2)
- x-intercepts are (−6, 0) and (2, 0)
- Axis of Symmetry is x = −2
- Maximum is 4 at x = −2
- Domain: All Real Numbers ℝ or (−∞, ∞)
- Range: y ≤ 4 or (−∞, 4]
- All Quadrants
- Increasing Interval (−∞, −2)
- Decreasing Interval (−2, ∞)
- Parent Function is $f(x) = |x|$

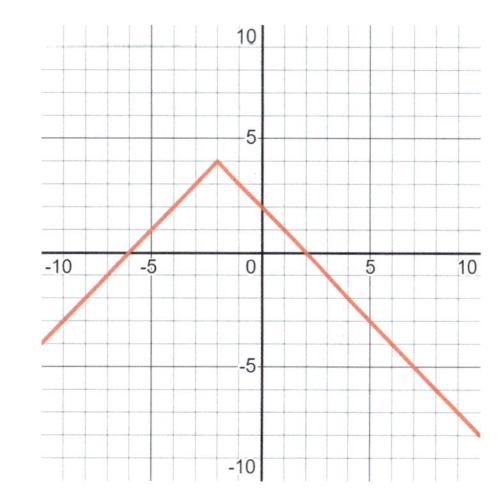

FOD #33

$$f(x) = -|x - 2| - 3$$

- Absolute Value Function
- Reflects over x-axis
- Horizontal shift right 2
- Vertical shift down 3
- Vertex is (2, −3)
- y-intercept is (0, −5)
- No x-intercepts
- Axis of Symmetry is x = 2
- Maximum is −3 at x = 2
- Domain: All Real Numbers ℝ or (−∞, ∞)
- Range: y ≤ −3 or (−∞, −3]
- Quadrants III & IV
- Increasing Interval (−∞, 2)
- Decreasing Interval (2, ∞)
- Parent Function is $f(x) = |x|$

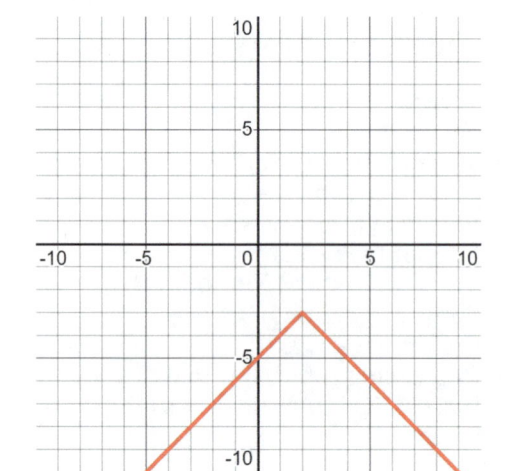

© Debra Richardson 2022

FOD #34

$$f(x) = -\frac{1}{2}|x + 4|$$

- Absolute Value Function
- Reflects over x-axis
- Vertical compression (wide) by a factor of ½
- Horizontal shift left 4
- Vertex is (−4, 0)
- y-intercept is (0, −2)
- x-intercept is (−4, 0)
- Axis of Symmetry is x = −4
- Maximum is 0 at x = − 4
- Domain: All Real Numbers ℝ or (−∞, ∞)
- Range: y ≤ 0 or (−∞, 0]
- Quadrants III & IV
- Increasing Interval (−∞, −4)
- Decreasing Interval (−4, ∞)
- Parent Function is $f(x) = |x|$

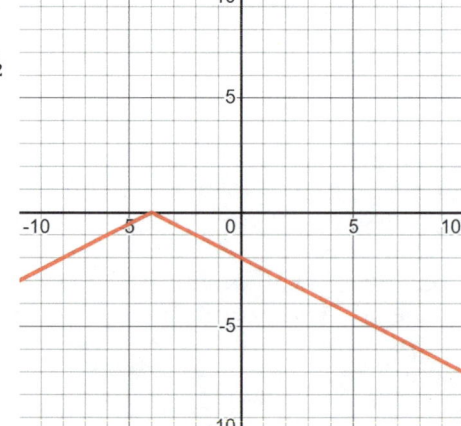

© Debra Richardson 2022

FOD #35

$$f(x) = 2|x - 3| - 4$$

- Absolute Value Function
- Vertical stretch (narrow) by a factor of 2
- Horizontal shift right 3
- Vertical shift down 4
- Vertex is (3, −4)
- y-intercept is (0, 2)
- x-intercepts are (1, 0) and (5, 0)
- Axis of Symmetry is x = 3
- Minimum is −4 at x = 3
- Domain: All Real Numbers $\mathbb{R}$ or (−∞, ∞)
- Range: y ≥ −4 or [−4, ∞)
- Quadrants I, II, & IV
- Decreasing Interval (−∞, 3)
- Increasing Interval (3, ∞)
- Parent Function is $f(x) = |x|$

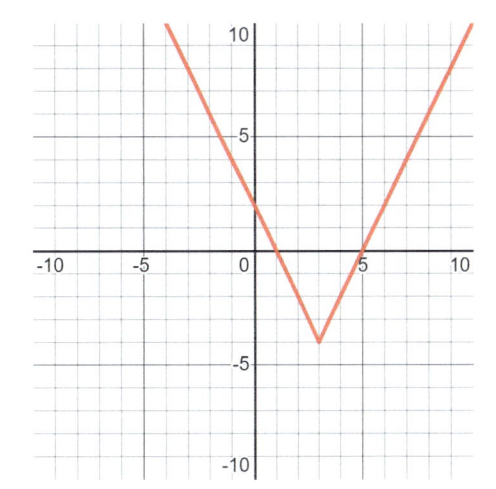

© Debra Richardson 2022

FOD #36

$$f(x) = -4|x - 2| + 4$$

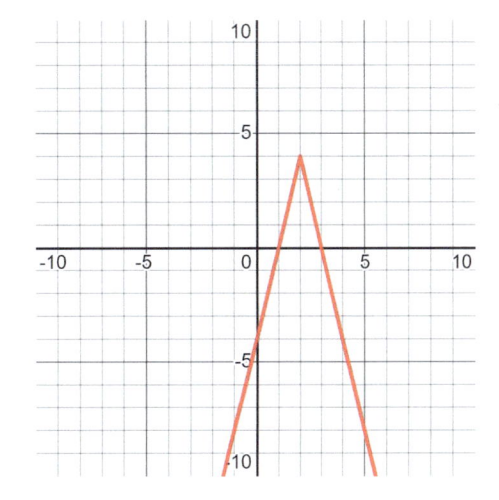

- Absolute Value Function
- Reflects over x-axis
- Vertical stretch (narrow) by a factor of 4
- Horizontal shift right 2
- Vertical shift up 4
- Vertex is (2, 4)
- y-intercept is (0, −4)
- x-intercepts are (1, 0) and (3, 0)
- Axis of Symmetry is x = 2
- Maximum is 4 at x = 2
- Domain: All Real Numbers $\mathbb{R}$ or (−∞, ∞)
- Range: y ≤ 4 or (−∞, 4]
- Quadrants I, III & IV
- Increasing Interval (−∞, 2)
- Decreasing Interval (2, ∞)
- Parent Function is $f(x) = |x|$

© Debra Richardson 2022

FOD #37

$$f(x) = \sqrt{x}$$

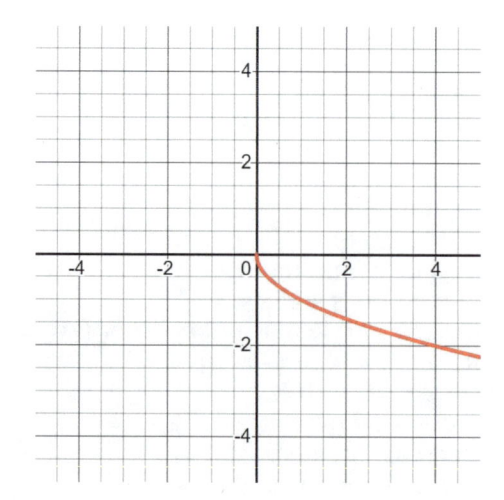

- Square Root Parent Function
- Radical Function
- Relation
- Continuous
- Nonlinear
- No symmetry
- Increasing
- Starting point is (0, 0)
- y-intercept is (0, 0)
- x-intercept is (0, 0)
- Minimum is 0 at x = 0
- Domain: x ≥ 0 or [0, ∞)
- Range: y ≥ 0 or [0, ∞)
- Quadrant I

© Debra Richardson 2022

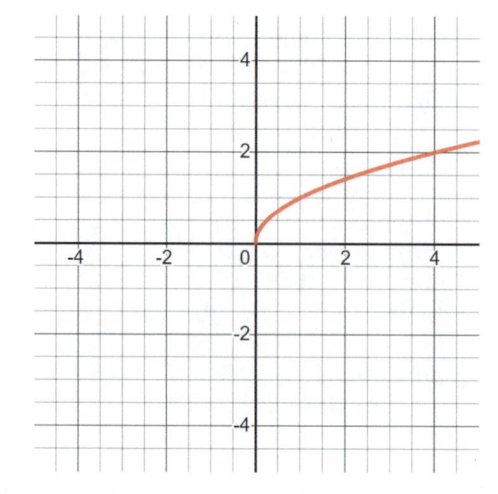

FOD #38

$$f(x) = -\sqrt{x}$$

- Square Root Function
- Radical Function
- Reflects over x-axis
- Relation, Continuous, Nonlinear
- No symmetry
- Decreasing
- Starting point is (0, 0)
- y-intercept is (0, 0)
- x-intercept is (0, 0)
- Maximum is 0 at x = 0
- Domain: x ≥ 0 or [0, ∞)
- Range: y ≤ 0 or (–∞, 0]
- Quadrant IV
- Parent Function is $f(x) = \sqrt{x}$

© Debra Richardson 2022

FOD #39

$$f(x) = \sqrt{x-3}$$

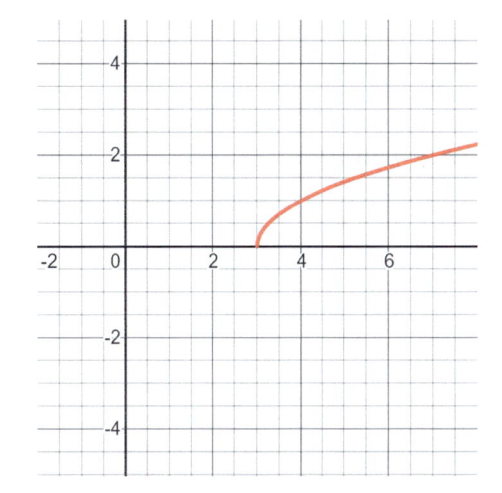

- Square Root Function
- Radical Function
- Relation, Continuous, Nonlinear
- No symmetry
- Increasing
- Horizontal shift right 3
- Starting point is (3, 0)
- No y-intercept
- x-intercept is (3, 0)
- Minimum is 0 at x = 3
- Domain: x ≥ 3 or [3, ∞)
- Range: y ≥ 0 or [0, ∞)
- Quadrant I
- Parent Function is $f(x) = \sqrt{x}$

© Debra Richardson 2022

FOD #40

$$f(x) = \sqrt{x} + 2$$

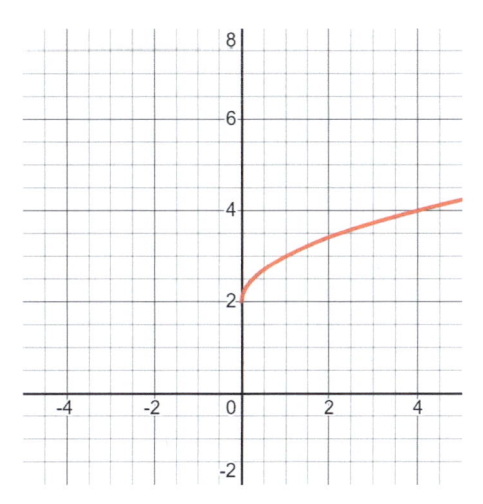

- Square Root Function
- Radical Function
- Relation, Continuous, Nonlinear
- No symmetry
- Increasing
- Vertical shift up 2
- Starting point is (0, 2)
- No x-intercept
- y-intercept is (0, 2)
- Minimum is 2 at x = 0
- Domain: x ≥ 0 or [0, ∞)
- Range: y ≥ 2 or [2, ∞)
- Quadrant I
- Parent Function is $f(x) = \sqrt{x}$

© Debra Richardson 2022

FOD #41

$$f(x) = -\sqrt{x} - 5$$

- Square Root Function
- Radical Function
- Reflects over x-axis
- Relation, Continuous, Nonlinear
- No symmetry
- Decreasing
- Vertical shift down 5
- Starting point is (0, −5)
- y-intercept is (0, −5)
- No x-intercept
- Maximum is −5 at x = 0
- Domain: x ≥ 0 or [0, ∞)
- Range: y ≤ −5 or (−∞, −5]
- Quadrant IV
- Parent Function is $f(x) = \sqrt{x}$

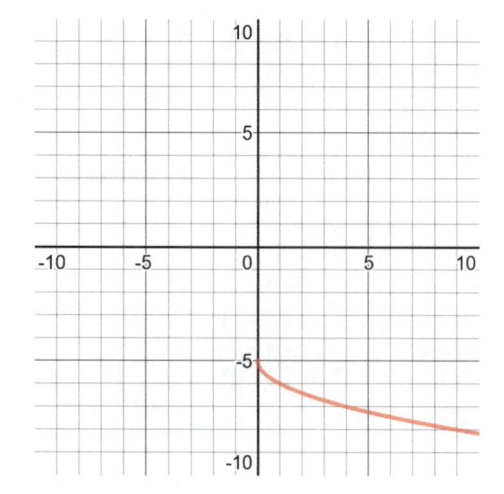

© Debra Richardson 2022

FOD #42

$$f(x) = \sqrt{x+1} - 1$$

- Square Root Function
- Radical Function
- Relation, Continuous, Nonlinear
- No symmetry
- Increasing
- Crosses the origin
- Horizontal shift left 1
- Vertical shift down 1
- Starting point is (−1, −1)
- x-intercept is (0, 0)
- y-intercept is (0, 0)
- Minimum is −1 at x = −1
- Domain: x ≥ −1 or [−1, ∞)
- Range: y ≥ −1 or [−1, ∞)
- Quadrants I & III
- Parent Function is $f(x) = \sqrt{x}$

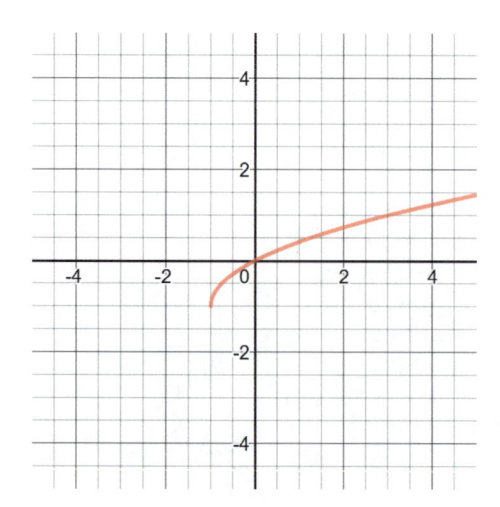

© Debra Richardson 2022

FOD #43

$$f(x) = -\sqrt{x+4} + 1$$

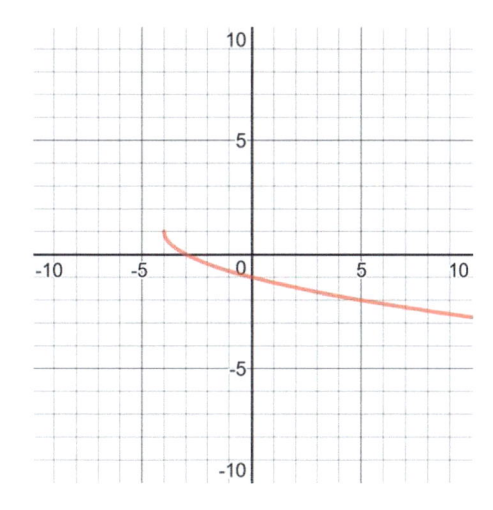

- Square Root Function
- Radical Function
- Reflects over x-axis
- Relation, Continuous, Nonlinear
- No symmetry
- Decreasing
- Horizontal shift left 4
- Vertical shift up 1
- Starting point is (−4, 1)
- y-intercept is (0, −1)
- x-intercept is (−3, 0)
- Maximum is 1 at x = −4
- Domain: x ≥ −4 or [−4, ∞)
- Range: y ≤ 1 or (−∞, 1]
- Quadrants II, III & IV
- Parent Function is $f(x) = \sqrt{x}$

© Debra Richardson 2022

FOD #44

$$f(x) = 4\sqrt{x} - 4$$

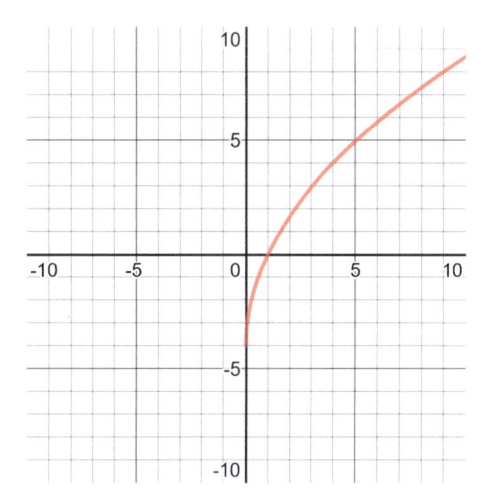

- Squarc Root Function
- Radical Function
- Relation, Continuous, Nonlinear
- No symmetry
- Increasing function
- Vertical stretch by a factor of 4
- Vertical shift down 4
- Starting point is (0, −4)
- x-intercept is (1, 0)
- y-intercept is (0, −4)
- Minimum is −4 at x = 0
- Domain: x ≥ 0 or [0, ∞)
- Range: y ≥ −4 or [−4, ∞)
- Quadrants I & IV
- Parent Function is $f(x) = \sqrt{x}$

© Debra Richardson 2022

FOD #45

$$f(x) = x^3$$

- Cubic Parent Function
- Degree 3 Polynomial
- Relation, Continuous, Nonlinear
- Symmetry across the origin
- Odd function
- Increasing
- No maximum, no minimum
- y-intercept is (0, 0)
- x-intercept is (0, 0)
- Domain: All Real Numbers $\mathbb{R}$ or (−∞, ∞)
- Range: All Real Numbers $\mathbb{R}$ or (−∞, ∞)
- Quadrants I & III
- End Behavior: As x → −∞, f(x) → −∞
- End Behavior: As x → +∞, f(x) → +∞

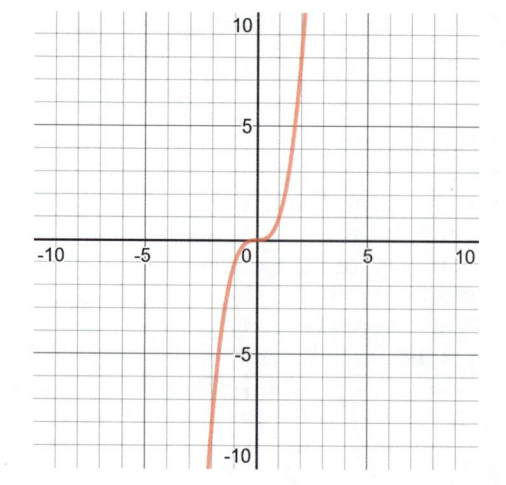

© Debra Richardson 2022

FOD #46

$$f(x) = -x^3$$

- Cubic Function
- Degree 3 Polynomial
- Reflects over x-axis
- Symmetry across the origin
- Odd function
- Decreasing
- No maximum, no minimum
- y-intercept is (0, 0)
- x-intercept is (0, 0)
- Domain: All Real Numbers $\mathbb{R}$ or (−∞, ∞)
- Range: All Real Numbers $\mathbb{R}$ or (−∞, ∞)
- Quadrants II & IV
- Parent function is $f(x) = x^3$
- End Behavior: As x → −∞, f(x) → +∞
- End Behavior: As x → +∞, f(x) → −∞

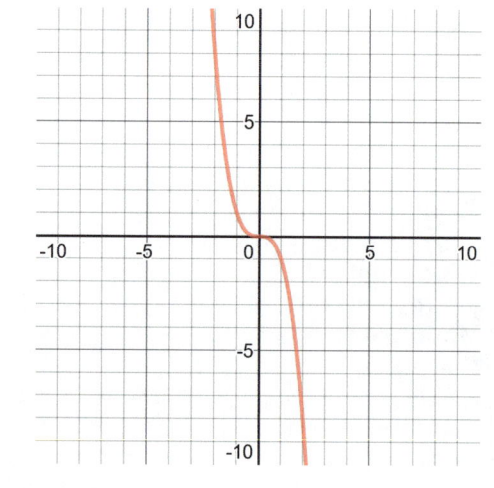

© Debra Richardson 2022

FOD #47

$$f(x) = -x^3 + 8$$

- Cubic Function
- Degree 3 Polynomial
- Reflects over x-axis
- Decreasing
- Vertical shift up 8
- No maximum, no minimum
- y-intercept is (0, 8)
- x-intercept is (2, 0)
- Domain: All Real Numbers $\mathbb{R}$ or (–∞, ∞)
- Range: All Real Numbers $\mathbb{R}$ or (–∞, ∞)
- Quadrants I, II & IV
- Parent function is $f(x) = x^3$
- End Behavior: As x→ –∞, f(x)→ +∞
- End Behavior: As x→ +∞, f(x)→ –∞

© Debra Richardson 2022

FOD #48

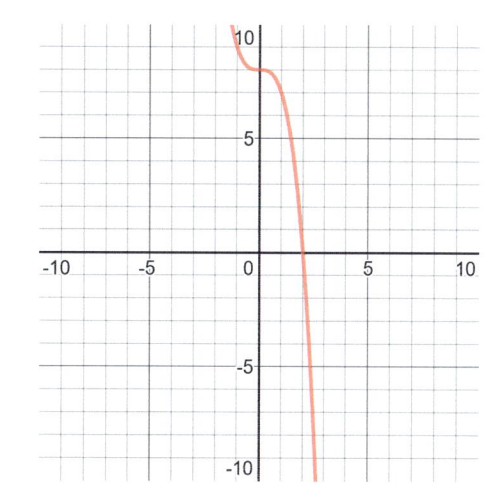

$$f(x) = \frac{1}{2}x^3 - 4$$

- Cubic Function
- Degree 3 Polynomial
- Vertical shift down 4
- Vertical compression (wide) by a factor of ½
- Increasing
- No maximum, no minimum
- y-intercept is (0, –4)
- x-intercept is (2, 0)
- Domain: All Real Numbers $\mathbb{R}$ or (–∞, ∞)
- Range: All Real Numbers $\mathbb{R}$ or (–∞, ∞)
- Quadrants I, III & IV
- Parent function is $f(x) = x^3$
- End Behavior: As x → –∞, f(x) → –∞
- End Behavior: As x → +∞, f(x) → +∞

© Debra Richardson 2022

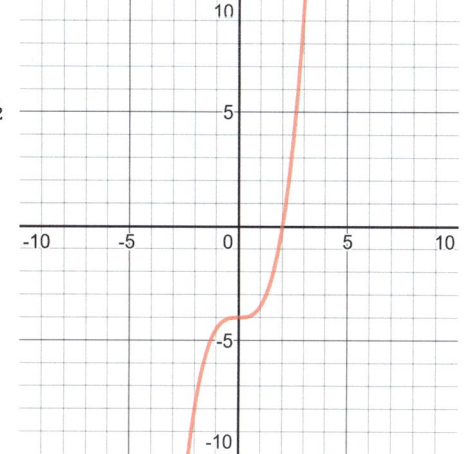

FOD #49 $f(x) = 4(x-2)^3$

- Cubic Function
- Degree 3 Polynomial
- Horizontal shift right 2
- Vertical stretch (narrow) by a factor of 4
- Increasing
- No maximum, no minimum
- y-intercept is (0, –32)
- x-intercept is (2, 0)
- Domain: All Real Numbers $\mathbb{R}$ or (–∞, ∞)
- Range: All Real Numbers $\mathbb{R}$ or (–∞, ∞)
- Quadrants I, III & IV
- Parent function is $f(x) = x^3$
- End Behavior: As x → –∞, f(x) → –∞
- End Behavior: As x → +∞, f(x) → +∞

© Debra Richardson 2022

FOD #50 $f(x) = x(x-2)(x+2)$

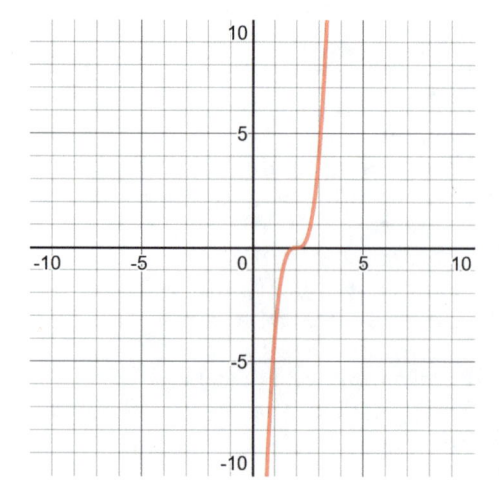

- Cubic Function
- Degree 3 Polynomial
- Equation in factored form
- Standard Form is $f(x) = x^3 - 4x$
- Increasing & Decreasing
- 2 turning points
- 1 relative maximum & 1 relative minimum
- y-intercept is (0, 0)
- x-intercepts are (–2, 0), (0, 0), (2, 0)
- Domain: All Real Numbers $\mathbb{R}$ or (–∞, ∞)
- Range: All Real Numbers $\mathbb{R}$ or (–∞, ∞)
- All Quadrants
- Parent function is $f(x) = x^3$
- End Behavior: As x → –∞, f(x) → –∞
- End Behavior: As x → +∞, f(x) → +∞

© Debra Richardson 2022

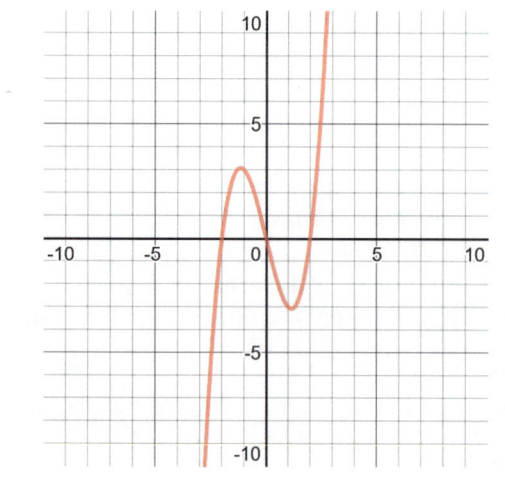

FOD #51

$$f(x) = -x(x-2)(x-4)$$

- Cubic Function
- Reflects over x-axis
- Equation in factored form
- Standard Form is $f(x) = -x^3 + 6x^2 - 8x$
- Decreasing & Increasing
- 2 turning points
- 1 relative minimum & 1 relative maximum
- y-intercept is (0, 0)
- x-intercepts are (0, 0), (2, 0), (4, 0)
- Domain: All Real Numbers $\mathbb{R}$ or (–∞, ∞)
- Range: All Real Numbers $\mathbb{R}$ or (–∞, ∞)
- Quadrants I, II & IV
- Parent function is $f(x) = x^3$
- End Behavior: As x → –∞, f(x) → +∞
- End Behavior: As x → +∞, f(x) → –∞

© Debra Richardson 2022

FOD #52

$$f(x) = x^2(x+3)$$

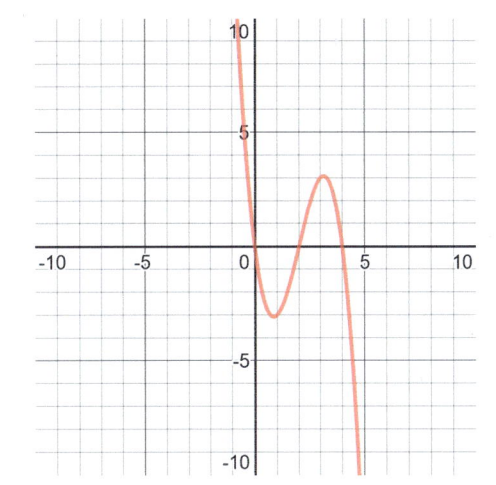

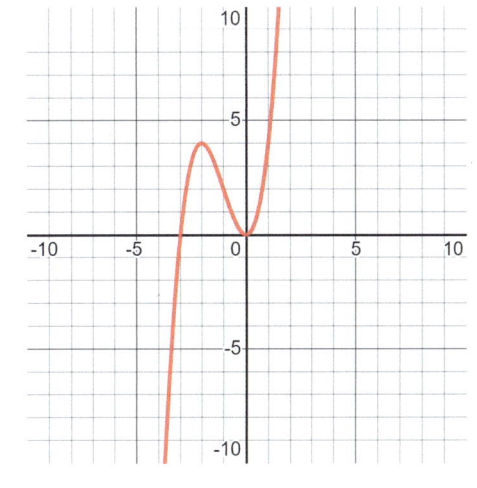

- Cubic Function
- Equation in factored form
- Standard Form is $f(x) = x^3 + 3x^2$
- 2 turning points
- Relative maximum at (–2, 4)
- Relative minimum at (0, 0)
- Increasing intervals (–∞, –2) and (0, ∞)
- Decreasing interval (–2, 0)
- y-intercept is (0, 0)
- x-intercepts are (–3, 0) and (0, 0)
- Domain: All Real Numbers $\mathbb{R}$ or (–∞, ∞)
- Range: All Real Numbers $\mathbb{R}$ or (–∞, ∞)
- Quadrants I, II & III
- Parent function is $f(x) = x^3$
- End Behavior: As x → –∞, f(x) → –∞
- End Behavior: As x → +∞, f(x) → +∞

© Debra Richardson 2022

FOD #53

$$f(x) = x^4$$

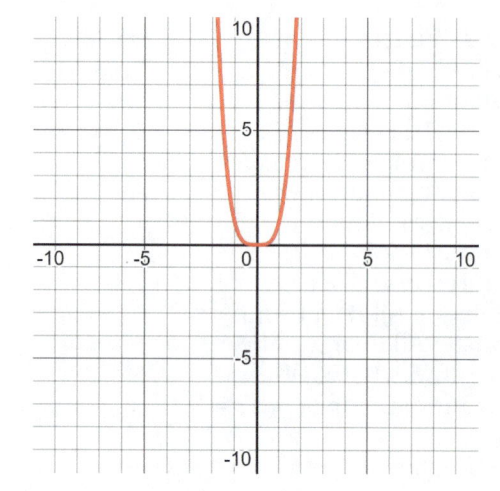

- Quartic Parent Function
- Degree 4 Polynomial
- U-shape, opens upward
- Touches the origin
- y-intercept is (0, 0)
- x-intercept is (0, 0)
- Symmetry over y-axis
- Even function
- Axis of Symmetry is x = 0
- Vertex is (0, 0)
- Minimum is 0 at x = 0
- Domain: All Real Numbers $\mathbb{R}$ or (−∞, ∞)
- Range: y ≥ 0 or [0, ∞)
- Quadrants I & II
- Decreasing Interval (−∞, 0)
- Increasing Interval (0, ∞)

FOD #54

$$f(x) = -(x+3)^4$$

- Quartic Function
- Degree 4 Polynomial
- Reflects over x-axis
- Opens downward
- Horizontal shift left 3
- y-intercept is (0, −81)
- x-intercept is (−3, 0)
- Vertex is (−3, 0)
- Maximum is 0 at x = −3
- Domain: All Real Numbers $\mathbb{R}$ or (−∞, ∞)
- Range: y ≤ 0 or (−∞, 0]
- Quadrants III & IV
- Parent Function is $f(x) = x^4$
- Increasing Interval (−∞, −3)
- Decreasing Interval (−3, ∞)

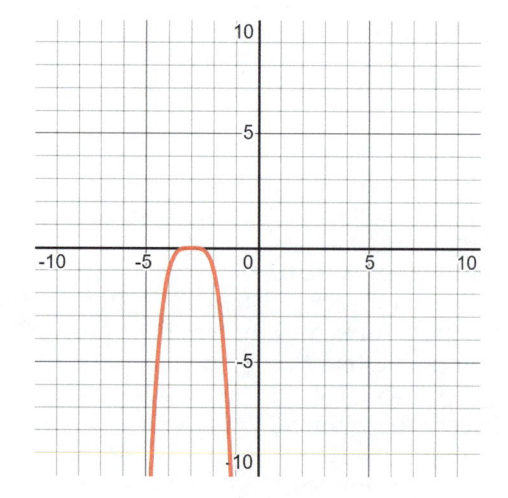

FOD #55 $\quad f(x) = x(x-1)(x-2)(x-3)$

- Quartic Function
- Degree 4 Polynomial
- Opens upward
- 3 turning points
- 1 relative maximum
- 2 relative minima
- y-intercept is (0, 0)
- x-intercepts are (0, 0), (1, 0), (2, 0), (3, 0)
- Domain: All Real Numbers $\mathbb{R}$ or (–∞, ∞)
- Range: y ≥ –1 or [–1, ∞)
- Quadrants I, II & IV
- Parent Function is $f(x) = x^4$
- End Behavior: As x → –∞, f(x) → +∞
- End Behavior: As x → +∞, f(x) → +∞

© Debra Richardson 2022

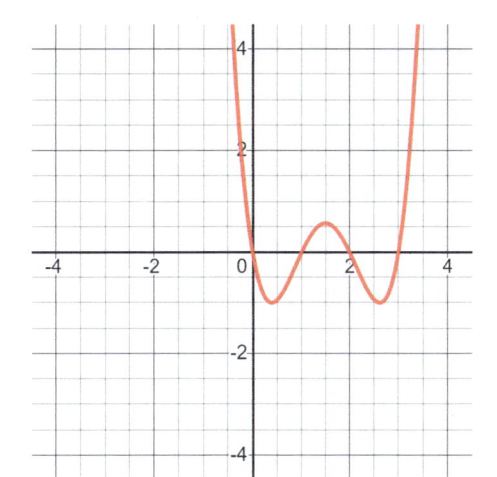

FOD #56 $\quad f(x) = -x(x+1)(x+2)(x+3)$

- Quartic Function
- Degree 4 Polynomial
- Negative leading coefficient
- Opens downward
- 3 turning points
- 2 relative maxima
- 1 relative minimum
- y-intercept is (0, 0)
- x-intercepts are (–3, 0), (–2, 0), (–1, 0), (0, 0)
- Domain: All Real Numbers $\mathbb{R}$ or (–∞, ∞)
- Range: y ≤ 1 or (–∞, 1]
- Quadrants II, III & IV
- Parent Function is $f(x) = x^4$
- End Behavior: As x → –∞, f(x) → –∞
- End Behavior: As x → +∞, f(x) → –∞

© Debra Richardson 2022

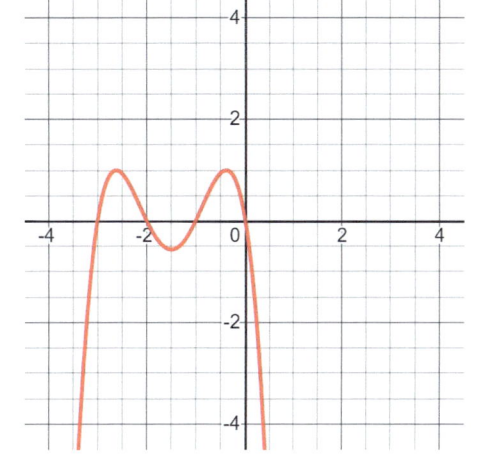

FOD #57 $f(x) = -x^2(x+3)(x-3)$

- Quartic Function
- Degree 4 Polynomial
- Negative leading coefficient
- Opens downward
- 3 turning points
- 2 relative maxima
- 1 relative minimum
- y-intercept is (0, 0)
- x-intercepts are (−3, 0), (0, 0), (3,0)
- Double root at (0, 0)
- Domain: All Real Numbers $\mathbb{R}$ or (−∞, ∞)
- Range: y ≤ 20.25 or (−∞, 20.25]
- All Quadrants
- Parent Function is $f(x) = x^4$
- End Behavior: As x → −∞, f(x) → −∞
- End Behavior: As x → +∞, f(x) → −∞

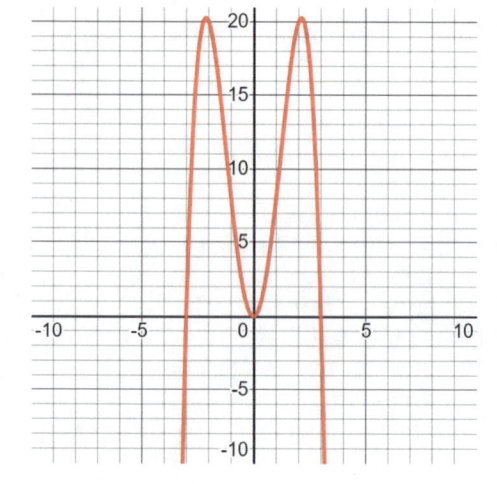

© Debra Richardson 2022

FOD #58 $f(x) = \dfrac{1}{x}$

- Rational Parent Function
- 2 curved pieces
- Each piece is continuous
- Relation
- Nonlinear
- Decreasing
- Does not touch or cross origin
- Does not touch or cross x-axis or y-axis
- No y-intercept
- No x-intercept
- Horizontal Asymptote y = 0
- Vertical Asymptote x = 0
- Domain: {x ∈ $\mathbb{R}$ | x ≠ 0} or (−∞, 0) U (0, ∞)
- Range: {y ∈ $\mathbb{R}$ | y ≠ 0} or (−∞, 0) U (0, ∞)
- Quadrants I & III

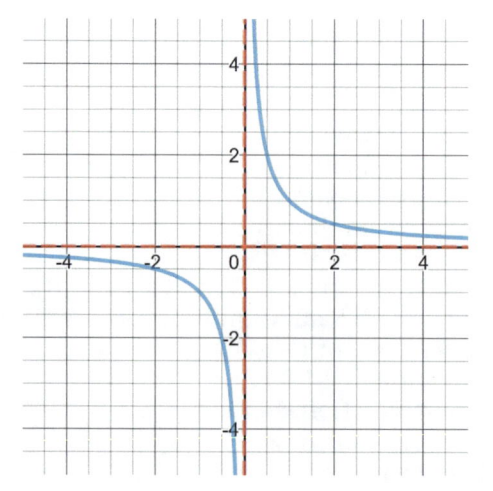

© Debra Richardson 2022

FOD #59

$$f(x) = -\frac{1}{x}$$

- Rational Function
- Reflects over x-axis
- 2 curved pieces
- Each piece is continuous
- Increasing
- Does not touch or cross origin
- Does not touch or cross x-axis or y-axis
- No y-intercept
- No x-intercept
- Horizontal Asymptote y = 0
- Vertical Asymptote x = 0
- Domain: {x ϵ ℝ | x ≠ 0} or (−∞, 0) U (0, ∞)
- Range: {y ϵ ℝ | y ≠ 0} or (−∞, 0) U (0, ∞)
- Quadrants II & IV
- Parent function is $f(x) = \frac{1}{x}$

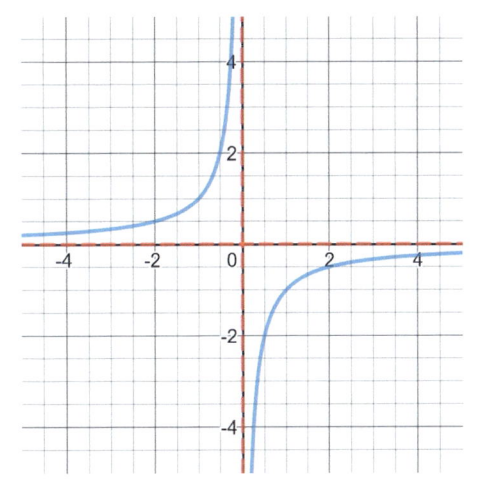

FOD #60

$$f(x) = \frac{1}{x+2}$$

- Rational Function
- 2 curved pieces
- Each piece is continuous
- Decreasing
- Horizontal shift left 2
- y-intercept is (0, 0.5)
- No x-intercept
- Horizontal Asymptote y = 0
- Vertical Asymptote x = −2
- Domain: { x ϵ ℝ | x ≠ −2 } or (−∞, −2) U (−2, ∞)
- Range: { y ϵ ℝ | y ≠ 0 } or (−∞, 0) U (0, ∞)
- Quadrants I, II & III
- Parent function is $f(x) = \frac{1}{x}$

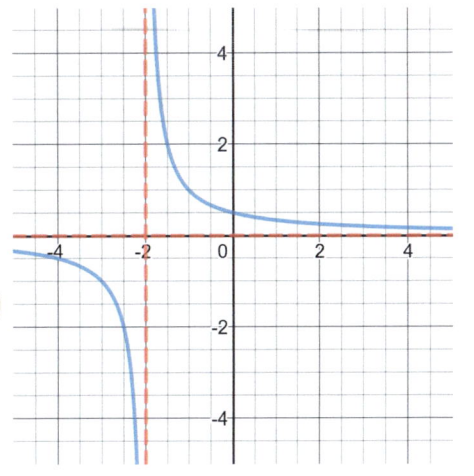

FOD #61

$$f(x) = \frac{1}{x-2}$$

- Rational Function
- 2 curved pieces
- Each piece is continuous
- Decreasing
- Horizontal shift right 2
- y-intercept is (0, –0.5)
- No x-intercept
- Horizontal Asymptote y = 0
- Vertical Asymptote x = 2
- Domain: {x ∈ ℝ | $x \neq 2$} or (–∞, 2) U (2, ∞)
- Range: {y ∈ ℝ | $y \neq 0$} or (–∞, 0) U (0, ∞)
- Quadrants I, III & IV
- Parent function is $f(x) = \frac{1}{x}$

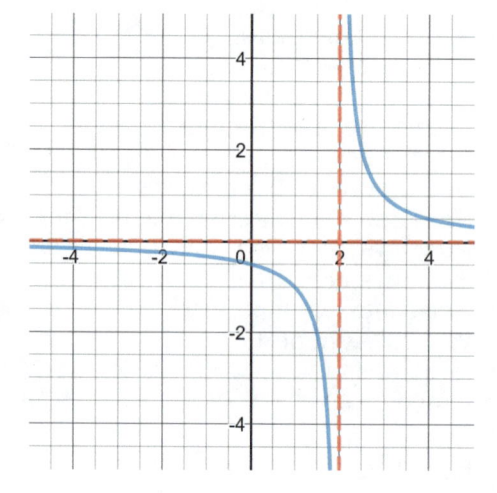

© Debra Richardson 2022

FOD #62

$$f(x) = \frac{1}{x} + 1$$

- Rational Function
- 2 curved pieces
- Each piece is continuous
- Decreasing
- Vertical shift up 1
- No y-intercept
- x-intercept is (–1, 0)
- Horizontal Asymptote y = 1
- Vertical Asymptote x = 0
- Domain: {x ∈ ℝ | $x \neq 0$} or (–∞, 0) U (0, ∞)
- Range: {y ∈ ℝ | $y \neq 1$} or (–∞, 1) U (1, ∞)
- Quadrants I, II & III
- Parent function is $f(x) = \frac{1}{x}$

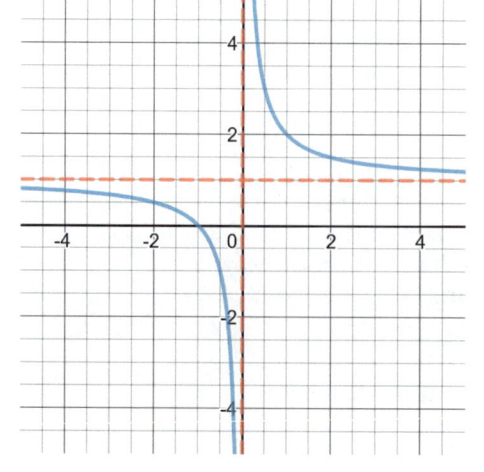

© Debra Richardson 2022

FOD #63

$$f(x) = \frac{1}{x} - 1$$

- Rational Function
- 2 curved pieces
- Each piece is continuous
- Decreasing
- Vertical shift down 1
- No y-intercept
- x-intercept is (1, 0)
- Horizontal Asymptote y = –1
- Vertical Asymptote x = 0
- Domain: {x ∈ ℝ | x ≠ 0} or (–∞, 0) U (0, ∞)
- Range: {y ∈ ℝ | y ≠ –1} or (–∞, –1) U (–1, ∞)
- Quadrants I, III & IV
- Parent function is $f(x) = \frac{1}{x}$

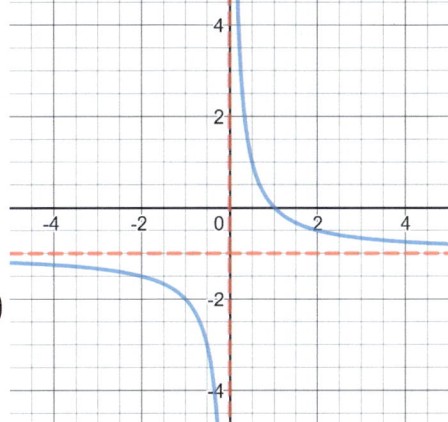

FOD #64

$$f(x) = \frac{1}{x+2} - 4$$

- Rational Function
- 2 curved pieces
- Each piece is continuous
- Decreasing
- Horizontal shift left 2
- Vertical shift down 4
- y-intercept is (0, –3.5)
- x-intercept is (–1.75, 0)
- Horizontal Asymptote y = –4
- Vertical Asymptote x = –2
- Domain: {x ∈ ℝ | x ≠ –2} or (–∞, –2) U (–2, ∞)
- Range: {y ∈ ℝ | y ≠ –4} or (–∞, –4) U (–4, ∞)
- Quadrants II, III & IV
- Parent function is $f(x) = \frac{1}{x}$

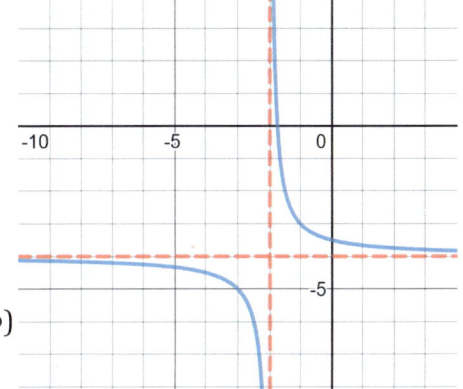

FOD #65

$$f(x) = -\frac{1}{x-2} + 4$$

- Rational Function
- Reflects over x-axis
- 2 curved pieces
- Each piece is continuous
- Increasing
- Horizontal shift right 2
- Vertical shift up 4
- y-intercept is (0, 4.5)
- x-intercept is (2.25, 0)
- Horizontal Asymptote y = 4
- Vertical Asymptote x = 2
- Domain: {x ∈ ℝ | x ≠ 2} or (−∞, 2) U (2, ∞)
- Range: {y ∈ ℝ | y ≠ 4} or (−∞, 4) U (4, ∞)
- Quadrants I, II & IV
- Parent function is $f(x) = \frac{1}{x}$

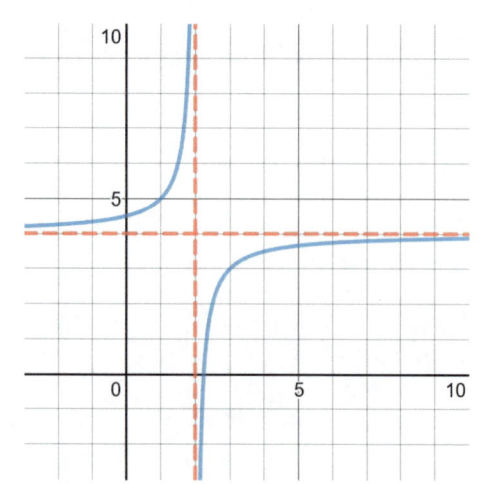

© Debra Richardson 2022

FOD #66

$$f(x) = \frac{x}{x(x-3)}$$

- Rational Function
- Simplified form of equation is $f(x) = \frac{1}{x-3}$
- Removable Discontinuity (hole) at $(0, -\frac{1}{3})$
- Decreasing
- Nonlinear
- Horizontal shift right 3
- No y-intercept
- No x-intercept
- Horizontal Asymptote y = 0
- Vertical Asymptote x = 3
- Domain: {x ∈ ℝ | x ≠ 0, 3} or (−∞, 0) U (0, 3) U (3, ∞)
- Range: {y ∈ ℝ | y ≠ −$\frac{1}{3}$, 0} or (−∞, −$\frac{1}{3}$) U (−$\frac{1}{3}$, 0) U (0, ∞)
- Quadrants I, III & IV
- Parent function is $f(x) = \frac{1}{x}$

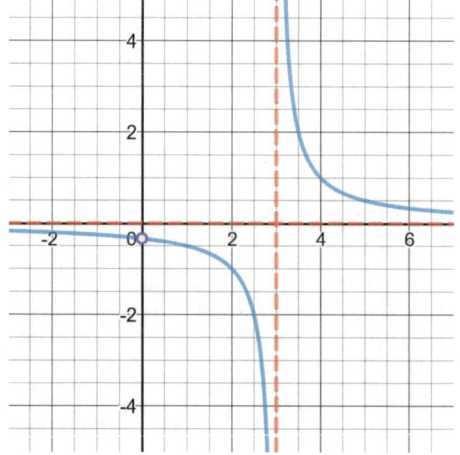

© Debra Richardson 2022

FOD #67

$$f(x) = \frac{x-1}{(x-1)(x-5)}$$

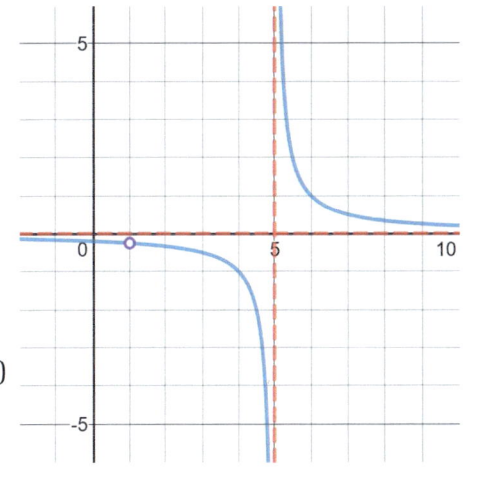

- Rational Function
- Simplified form of equation is $f(x) = \frac{1}{x-5}$
- Removable Discontinuity (hole) at $(1, -\frac{1}{4})$
- Decreasing
- Nonlinear
- Horizontal shift right 5
- y-intercept is $(0, -\frac{1}{5})$
- No x-intercept
- Horizontal Asymptote y = 0
- Vertical Asymptote x = 5
- Domain: {x ∈ ℝ | x ≠ 1, 5} or (-∞, 1) U (1, 5) U (5, ∞)
- Range: {y ∈ ℝ | y ≠ -$\frac{1}{4}$, 0} or (-∞, -$\frac{1}{4}$) U (-$\frac{1}{4}$, 0) U (0, ∞)
- Quadrants I, III & IV
- Parent function is $f(x) = \frac{1}{x}$

© Debra Richardson 2022

FOD #68

$$f(x) = \frac{x-2}{x^2-4}$$

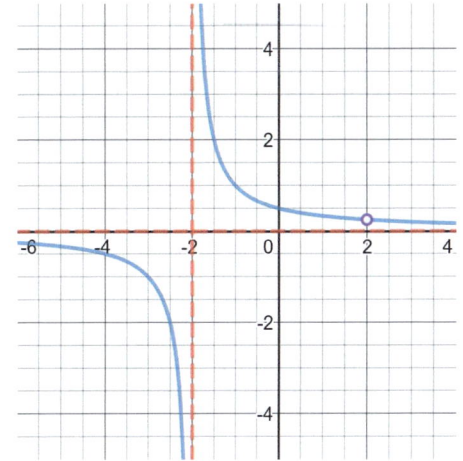

- Rational Function
- Factored form of equation $f(x) = \frac{x-2}{(x-2)(x+2)}$
- Simplified form of equation $f(x) = \frac{1}{x+2}$
- Decreasing
- Removable Discontinuity (hole) at $(2, \frac{1}{4})$
- Horizontal shift left 2
- y-intercept is $(0, \frac{1}{2})$
- No x-intercept
- Horizontal Asymptote y = 0
- Vertical Asymptote x = -2
- Domain: {x ∈ ℝ | x ≠ -2, 2} or (-∞, -2) U (-2, 2) U (2, ∞)
- Range: {y ∈ ℝ | y ≠ 0, $\frac{1}{4}$} or (-∞, 0) U (0, $\frac{1}{4}$) U ($\frac{1}{4}$, ∞)
- Quadrants I, II & III
- Parent function is $f(x) = \frac{1}{x}$

© Debra Richardson 2022

FOD #69

$$f(x) = \frac{x^2-5x+6}{x^2-7x+1}$$

- Rational Function
- Factored form of equation $f(x) = \frac{(x-3)(x-2)}{(x-4)(x-3)}$
- Simplified form of equation $f(x) = \frac{x-2}{x-4}$
- Horizontal shift right 4
- Vertical shift up 1
- Removable Discontinuity (hole) at (3, −1)
- Decreasing
- y-intercept is $(0, \frac{1}{2})$
- x-intercept is (2, 0)
- Horizontal Asymptote y = 1
- Vertical Asymptote x = 4
- Domain: $\{x \in \mathbb{R} \mid x \neq 3, 4\}$ or (−∞, 3) U (3, 4) U (4, ∞)
- Range: $\{y \in \mathbb{R} \mid y \neq -1, 1\}$ or (−∞, −1) U (−1, 1) U (1, ∞)
- Quadrants I, II & IV
- Parent function is $f(x) = \frac{1}{x}$

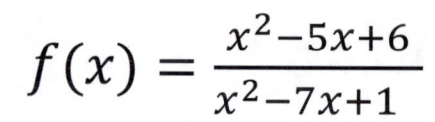

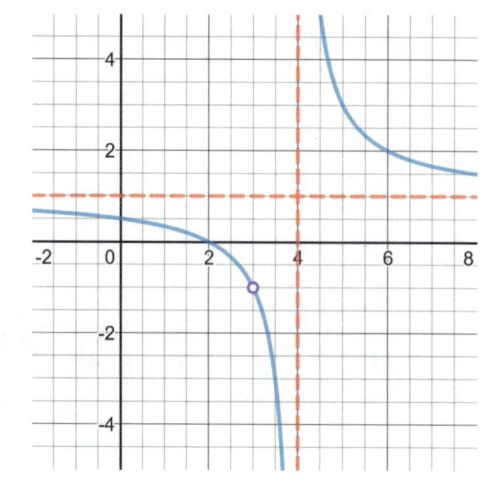

© Debra Richardson 2022

FOD #70

$$f(x) = 2^x$$

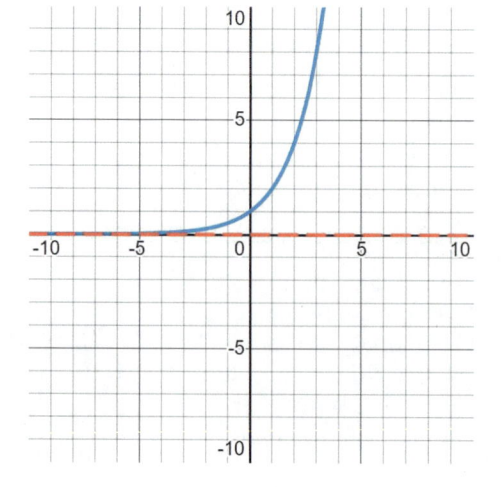

- Exponential Growth Function
- Base is 2
- Continuous
- Increasing
- y-intercept is (0, 1)
- No x-intercepts
- Horizontal Asymptote y = 0
- Domain: All Real Numbers $\mathbb{R}$ or (−∞, ∞)
- Range: y > 0 or (0, ∞)
- Quadrants I & II
- Parent Function is $f(x) = (b)^x$
- End Behavior: As x → −∞, f(x) → 0
- End Behavior: As x → +∞, f(x) → +∞

© Debra Richardson 2022

FOD #71

$$f(x) = \frac{1}{2}^x$$

- Exponential Decay Function
- Base is ½
- Continuous
- Decreasing
- y-intercept is (0, 1)
- No x-intercepts
- Horizontal Asymptote y = 0
- Domain: All Real Numbers $\mathbb{R}$ or (−∞, ∞)
- Range: y > 0 or (0, ∞)
- Quadrants I & II
- Parent Function is $f(x) = (b)^x$
- End Behavior: As x → −∞, f(x) → +∞
- End Behavior: As x → +∞, f(x) → 0

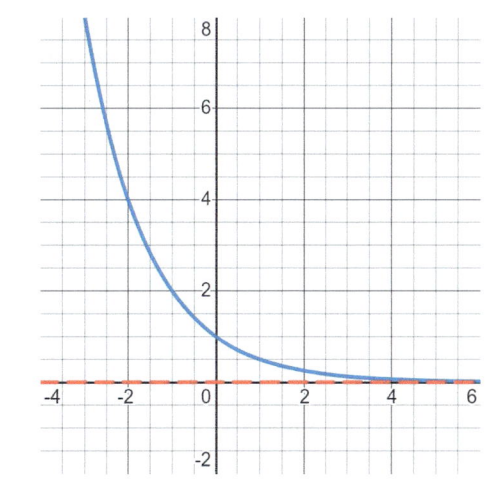

© Debra Richardson 2022

FOD #72

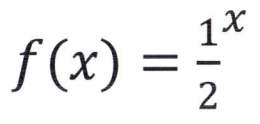

$$f(x) = 2^x + 3$$

- Exponential Growth Function
- Base is 2
- Continuous
- Increasing
- Vertical Shift up 3
- Horizontal Asymptote y = 3
- y-intercept is (0, 4)
- No x-intercepts
- Domain: All Real Numbers $\mathbb{R}$ or (−∞, ∞)
- Range: y > 3 or (3, ∞)
- Quadrants I & II
- Parent Function is $f(x) = (b)^x$
- End Behavior: As x → −∞, f(x) → 3
- End Behavior: As x → +∞, f(x) → +∞

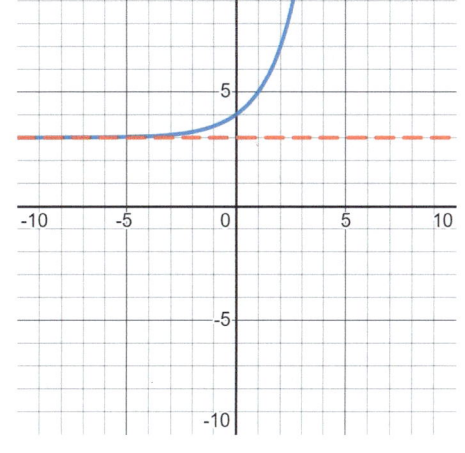

© Debra Richardson 2022

FOD #73

$$f(x) = 0.01(2)^x$$

- Exponential Growth Function
- Base is 2
- Increasing
- y-intercept is (0, 0.01)
- No x-intercepts
- Horizontal Asymptote y = 0
- Domain: All Real Numbers ℝ or (−∞, ∞)
- Range: y > 0 or (0, ∞)
- Quadrants I & II
- Parent Function is $f(x) = (b)^x$
- End Behavior: As x → −∞, f(x) → 0
- End Behavior: As x → +∞, f(x) → +∞
- Word Problem: A penny a day is doubled
- Initial Amount is $0.01

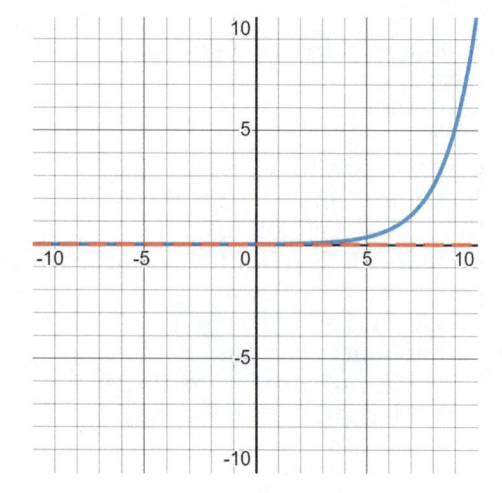

FOD #74

$$f(x) = -4(2)^x$$

- Exponential Growth Function
- Base is 2
- Growth is negative
- Reflects over x-axis
- Decreasing
- Vertical stretch by a factor of 4
- y-intercept is (0, −4)
- No x-intercepts
- Horizontal Asymptote y = 0
- Domain: All Real Numbers ℝ or (−∞, ∞)
- Range: y < 0 or (−∞, 0)
- Quadrants III & IV
- Parent Function is $f(x) = (b)^x$
- End Behavior: As x → −∞, f(x) → 0
- End Behavior: As x → +∞, f(x) → −∞

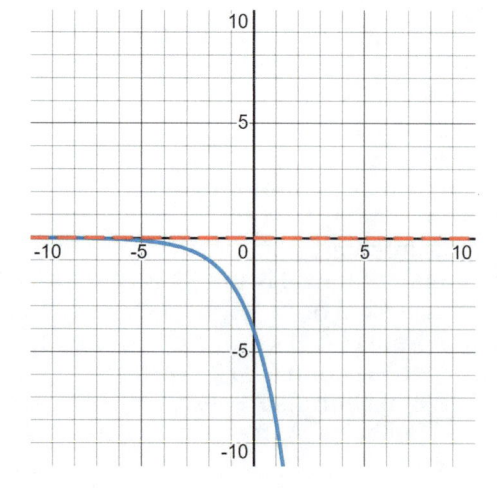

FOD #75

$$f(x) = -(2)^x - 4$$

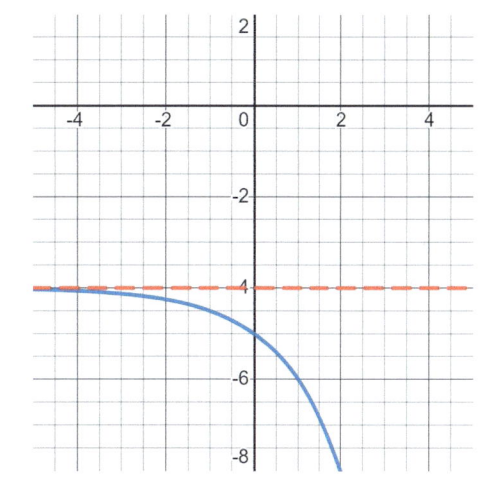

- Exponential Growth Function
- Base is 2
- Growth is negative
- Reflects over x-axis
- Decreasing
- Vertical shift down 4
- Horizontal Asymptote y = −4
- y-intercept is (0, −5)
- No x-intercepts
- Domain: All Real Numbers ℝ or (−∞, ∞)
- Range: y < −4 or (−∞, −4)
- Quadrants III & IV
- Parent Function is $f(x) = (b)^x$
- End Behavior: As x → −∞, f(x) → − 4
- End Behavior: As x → +∞, f(x) → − ∞

FOD #76

$$f(x) = 3^x + 2$$

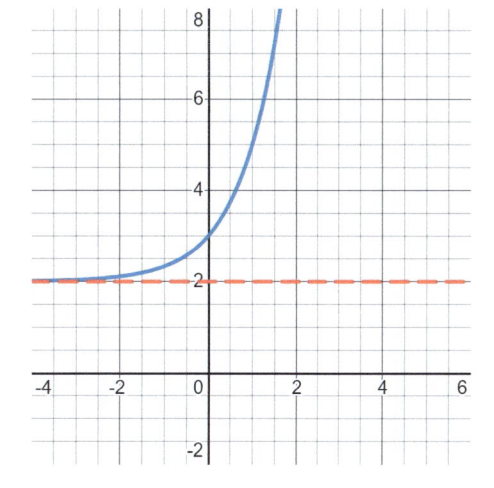

- Exponential Growth Function
- Base is 3
- Continuous
- Increasing
- Vertical Shift up 2
- Horizontal Asymptote y = 2
- y-intercept is (0, 3)
- No x-intercepts
- Domain: All Real Numbers ℝ or (−∞, ∞)
- Range: y > 2 or (2, ∞)
- Quadrants I & II
- Parent Function is $f(x) = (b)^x$
- End Behavior: As x → −∞, f(x) → 2
- End Behavior: As x → +∞, f(x) → +∞

FOD #77

$$f(x) = 3^{x-5} - 1$$

- Exponential Growth Function
- Base is 3
- Increasing
- Horizontal Shift right 5
- Vertical Shift down 1
- Horizontal Asymptote y = −1
- y-intercept is (0, −0.996)
- x-intercept is (5, 0)
- Domain: All Real Numbers $\mathbb{R}$ or (−∞, ∞)
- Range: y > −1 or (−1, ∞)
- Quadrants I, III & IV
- Parent Function is $f(x) = (b)^x$
- End Behavior: As x → −∞, f(x) → −1
- End Behavior: As x → +∞, f(x) → +∞

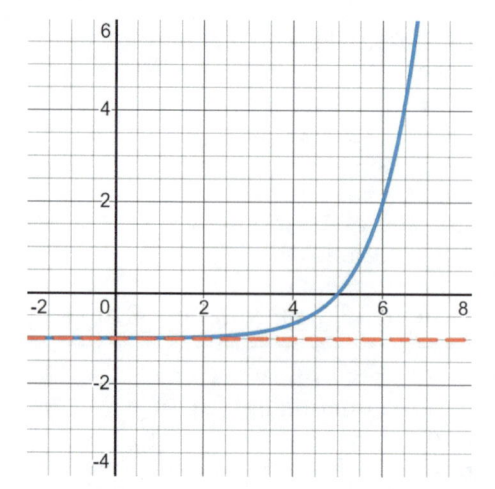

FOD #78

$$f(x) = 4^{-x} + 4$$

- Exponential Growth Function
- Base is 4
- Reflects over y-axis
- Continuous
- Decreasing
- Vertical shift up 4
- Horizontal Asymptote y = 4
- y-intercept is (0, 5)
- No x-intercepts
- Domain: All Real Numbers $\mathbb{R}$ or (−∞, ∞)
- Range: y > 4 or (4, ∞)
- Quadrants I & II
- Parent Function is $f(x) = (b)^x$
- End Behavior: As x → −∞, f(x) → +∞
- End Behavior: As x → +∞, f(x) → 4

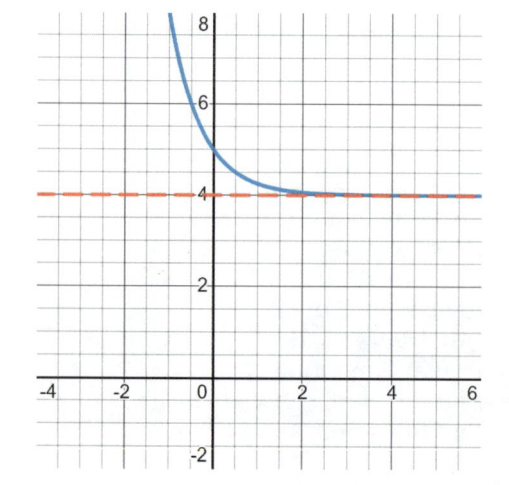

FOD #79 $f(x) = -(5)^x + 5$

- Exponential Growth Function
- Base is 5
- Growth is negative
- Reflects over x-axis
- Decreasing
- Vertical shift up 5
- Horizontal Asymptote y = 5
- y-intercept is (0, 4)
- x-intercept is (1, 0)
- Domain: All Real Numbers $\mathbb{R}$ or (−∞, ∞)
- Range: y < 5 or (−∞, 5)
- Quadrants I, II & IV
- Parent Function is $f(x) = (b)^x$
- End Behavior: As x → −∞, f(x) → 5
- End Behavior: As x → +∞, f(x) → −∞

FOD #80 $f(x) = 10^x$

- Exponential Growth Function
- Base is 10
- Continuous
- Increasing
- y-intercept is (0, 1)
- No x-intercepts
- Horizontal Asymptote y = 0
- Never touches x-axis
- Domain: All Real Numbers $\mathbb{R}$ or (−∞, ∞)
- Range: y > 0 or (0, ∞)
- Quadrants I & II
- Inverse is $f(x) = \log x$
- Parent Function is $f(x) = (b)^x$
- End Behavior: As $x \to -\infty, f(x) \to 0$
- End Behavior: As $x \to +\infty, f(x) \to +\infty$

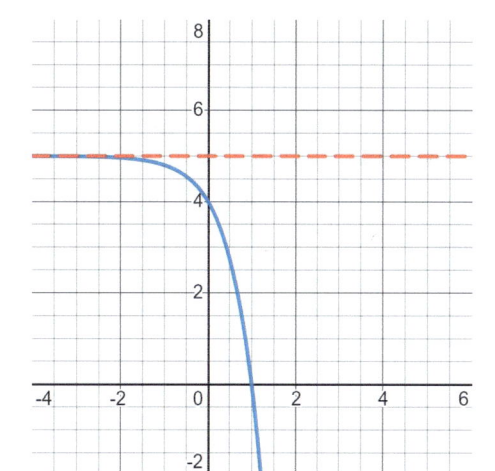

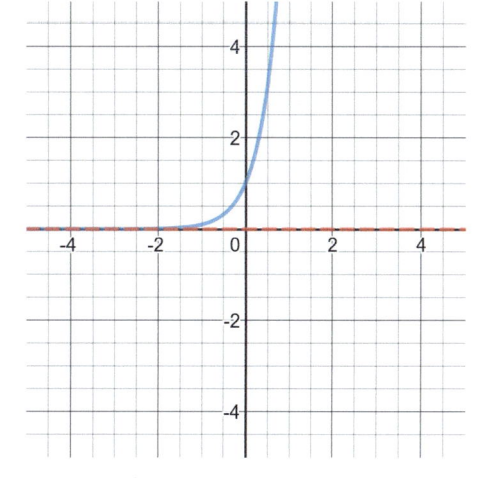

FOD #81

$$f(x) = \log x$$

- Logarithmic Parent Function
- Base of log is 10
- Inverse is $f(x) = 10^x$
- Relation, Continuous, Nonlinear
- Increasing
- x-intercept is (1, 0)
- No y-intercept
- Vertical Asymptote at x = 0
- No horizontal asymptote
- No minimum
- No maximum
- Domain: x > 0 or (0, ∞)
- Range: All Real Numbers $\mathbb{R}$ or (−∞, ∞)
- Quadrants I & IV

© Debra Richardson 2022

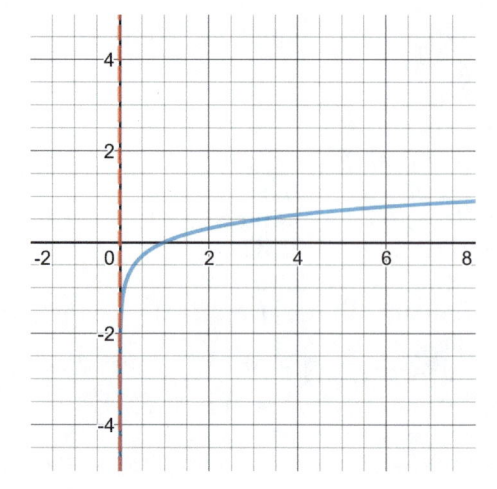

FOD #82

$$f(x) = \log(x - 4)$$

- Logarithmic Function
- Relation, Continuous, Nonlinear
- Increasing
- Horizontal shift right 4
- Vertical Asymptote at x = 4
- x-intercept is (5, 0)
- No y-intercept
- No minimum
- No maximum
- Domain: x > 4 or (4, ∞)
- Range: All Real Numbers $\mathbb{R}$ or (−∞, ∞)
- Quadrants I & IV
- Parent function is $f(x) = \log x$

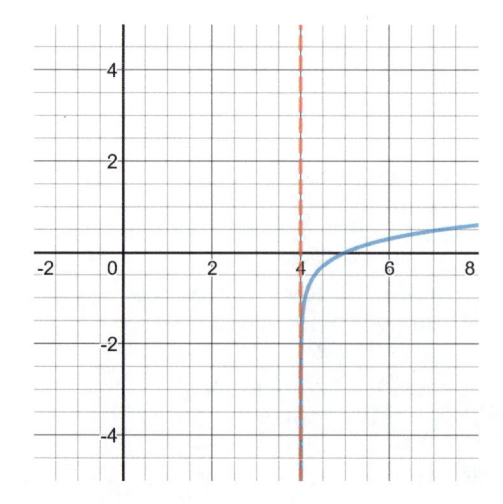

© Debra Richardson 2022

FOD #83

$$f(x) = \log(x + 5)$$

- Logarithmic Function
- Relation, Continuous, Nonlinear
- Increasing
- Horizontal shift left 5
- Vertical Asymptote at x = −5
- x-intercept is (−4, 0)
- y-intercept is (0, 0.7)
- No minimum
- No maximum
- Domain: x > −5 or (−5, ∞)
- Range: All Real Numbers $\mathbb{R}$ or (−∞, ∞)
- Quadrants I, II & III
- Parent function is $f(x) = \log x$

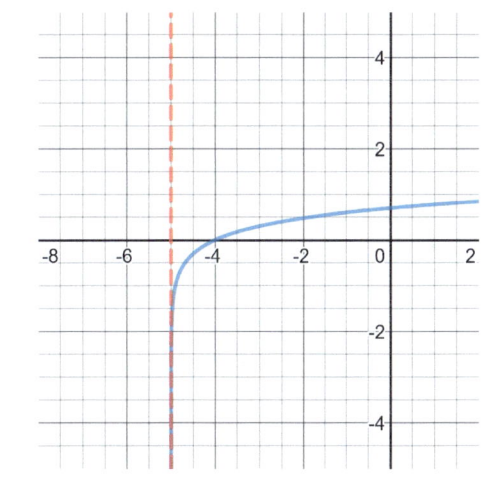

© Debra Richardson 2022

FOD #84

$$f(x) = \log(x + 1) - 1$$

- Logarithmic Function
- Relation, Continuous, Nonlinear
- Increasing
- Horizontal shift left 1
- Vertical Asymptote at x = −1
- Vertical shift down 1
- x-intercept is (9, 0)
- y-intercept is (0, −1)
- No minimum
- No maximum
- Domain: x > −1 or (−1, ∞)
- Range: All Real Numbers $\mathbb{R}$ or (−∞, ∞)
- Quadrants I, III & IV
- Parent function is $f(x) = \log x$

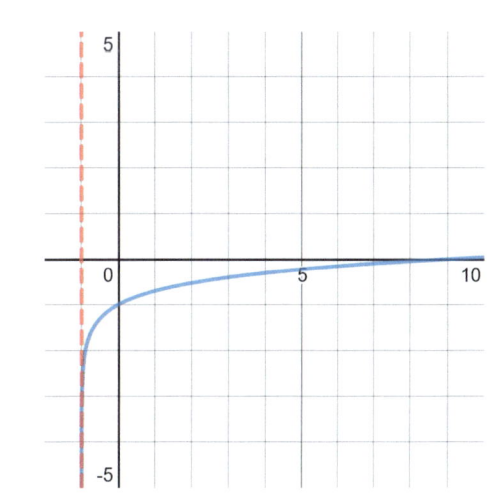

© Debra Richardson 2022

FOD #85

$$f(x) = -\log x$$

- Logarithmic Function
- Relation, Continuous, Nonlinear
- Negative
- Reflects over x-axis
- Decreasing
- x-intercept is (1, 0)
- No y-intercept
- Vertical Asymptote at x = 0
- No maximum
- No minimum
- Domain: x > 0 or (0, ∞)
- Range: All Real Numbers ℝ or (−∞, ∞)
- Quadrants I & IV
- Parent function is $f(x) = \log x$

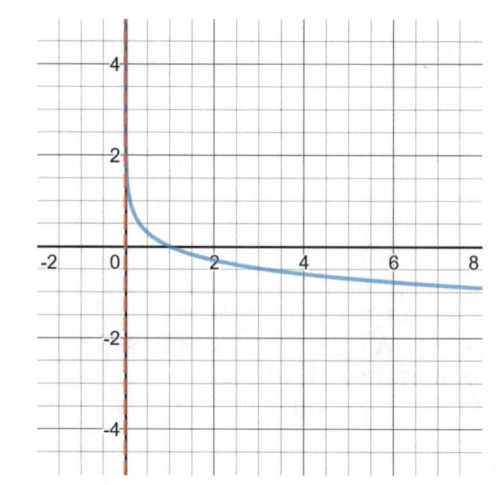

© Debra Richardson 2022

FOD #86

$$f(x) = -\log(x - 5)$$

- Logarithmic Function
- Reflects over x-axis
- Decreasing
- Horizontal shift right 5
- x-intercept is (6, 0)
- No y-intercept
- Vertical Asymptote at x = 5
- No maximum
- No minimum
- Domain: x > 5 or (5, ∞)
- Range: All Real Numbers ℝ or (−∞, ∞)
- Quadrants I & IV
- Parent function is $f(x) = \log x$

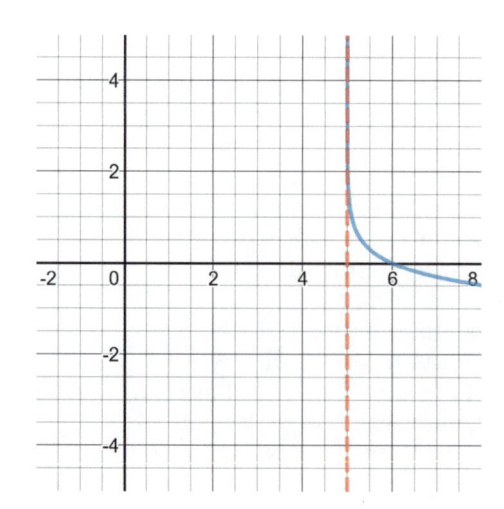

© Debra Richardson 2022

FOD #87 $f(x) = -\log(x + 6) + 1$

- Logarithmic Function
- Reflects over x-axis
- Decreasing
- Horizontal shift left 6
- Vertical Asymptote at x = −6
- Vertical shift up 1
- x-intercept is (4, 0)
- y-intercept is (0, 0.2218)
- No maximum
- No minimum
- Domain: x > −6 or (−6, ∞)
- Range: All Real Numbers ℝ or (−∞, ∞)
- Quadrants I, II, & IV
- Parent function is $f(x) = \log x$

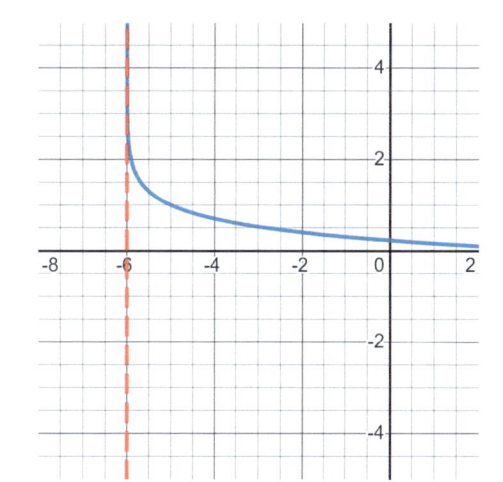

FOD #88 $f(x) = -2\log(x + 1)$

- Logarithmic Function
- Reflects over x-axis
- Decreasing
- Vertical stretch by a factor of 2
- Horizontal shift left 1
- Vertical Asymptote at x = −1
- x-intercept is (0, 0)
- y-intercept is (0, 0)
- No maximum, no minimum
- Domain: x > −1 or (−1, ∞)
- Range: All Real Numbers ℝ or (−∞, ∞)
- Quadrants II & IV
- Parent function is $f(x) = \log x$

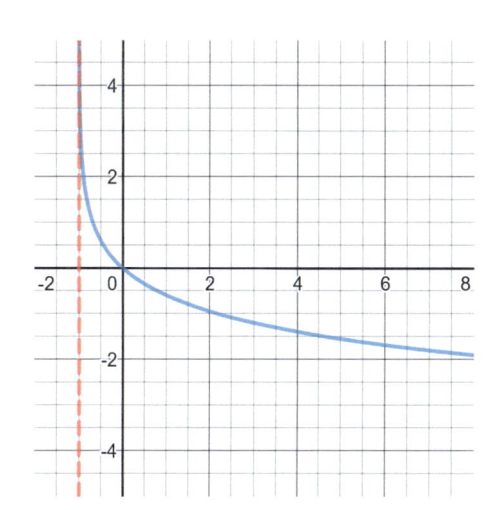

FOD #89

$$f(x) = e^x$$

- Exponential Growth Function
- Parent Function
- Base is "*e*"
- Euler's number $e \approx 2.7182818...$
- Continuous
- Increasing
- y-intercept is (0, 1)
- No x-intercepts
- Horizontal Asymptote y = 0
- Never touches x-axis
- No vertical asymptote
- Domain: All Real Numbers $\mathbb{R}$ or $(-\infty, \infty)$
- Range: y > 0 or $(0, \infty)$
- Quadrants I & II
- Inverse is $f(x) = \ln x$

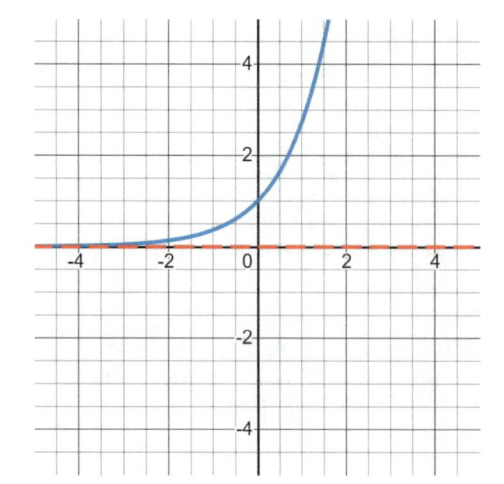

© Debra Richardson 2022

FOD #90

$$f(x) = \ln x$$

- Natural Logarithmic Function
- Base of natural log is "*e*"
- Inverse is $f(x) = e^x$
- Parent Function
- Continuous
- Increasing
- x-intercept is (1, 0)
- No y-intercept
- Vertical Asymptote at x = 0
- No horizontal asymptote
- No minimum
- No maximum
- Domain: x > 0 or $(0, \infty)$
- Range: All Real Numbers $\mathbb{R}$ or $(-\infty, \infty)$
- Quadrants I & IV

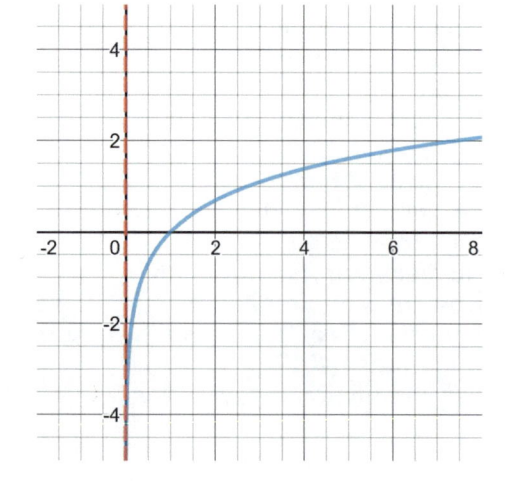

© Debra Richardson 2022

FOD #91 $\qquad$ $f(x) = e^x + 4$

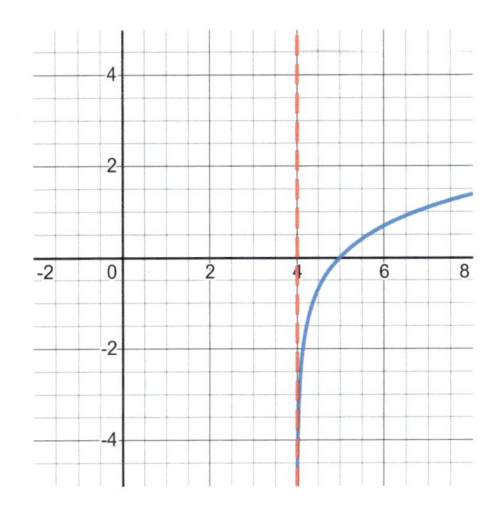

- Exponential Growth Function
- Base is "e"
- Continuous
- Increasing
- Horizontal shift up 4
- Horizontal Asymptote y = 4
- y-intercept is (0, 5)
- No x-intercepts
- Domain: All Real Numbers $\mathbb{R}$ or (−∞, ∞)
- Range: y > 4 or (4, ∞)
- Quadrants I & II
- Parent Function is $f(x) = e^x$

© Debra Richardson 2022

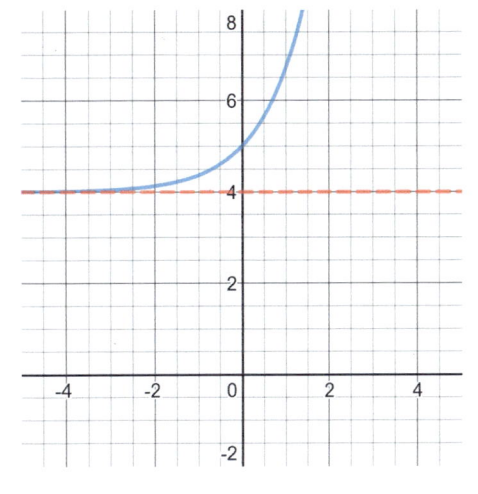

FOD #92 $\qquad$ $f(x) = \ln(x - 4)$

- Natural Logarithmic Function
- Inverse is $f(x) = e^x + 4$
- Continuous
- Increasing
- Horizontal shift right 4
- Vertical Asymptote at x = 4
- x-intercept is (5, 0)
- No y-intercept
- No minimum
- No maximum
- Domain: x > 4 or (4, ∞)
- Range: All Real Numbers $\mathbb{R}$ or (−∞, ∞)
- Quadrant I & IV
- Parent Function is $f(x) = \ln x$

© Debra Richardson 2022

FOD #93

$$f(x) = e^x - 3$$

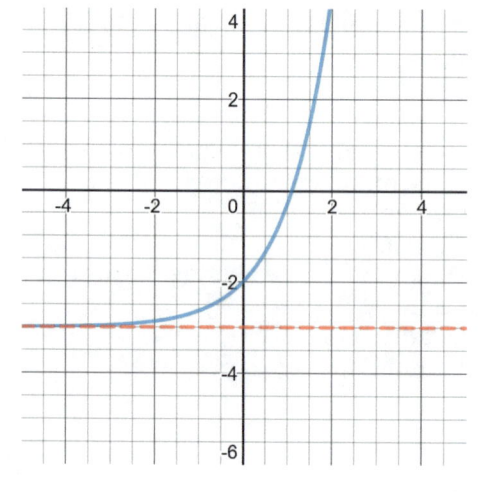

- Exponential Growth Function
- Base is "e"
- Continuous
- Increasing
- Vertical shift down 3
- Horizontal Asymptote y = −3
- y-intercept is (0, −2)
- x-intercept is (1.098, 0)
- Domain: All Real Numbers $\mathbb{R}$ or (−∞, ∞)
- Range: y > −3 or (−3, ∞)
- Quadrants I, III & IV
- Parent function is $f(x) = e^x$

FOD #94

$$f(x) = \ln(x + 3)$$

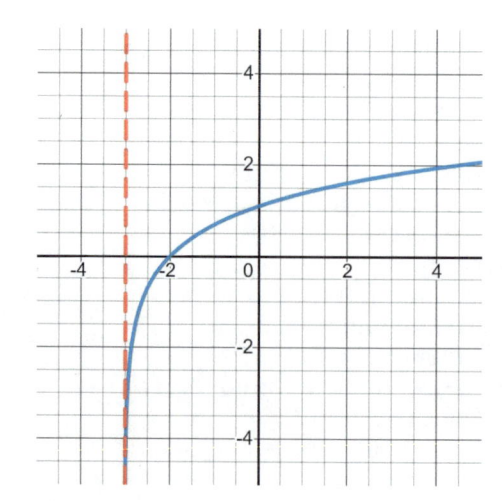

- Natural Logarithmic Function
- Inverse is $f(x) = e^x - 3$
- Continuous
- Increasing
- Horizontal shift left 3
- Vertical Asymptote at x = −3
- x-intercept is (−2, 0)
- y-intercept is (0, 1.099)
- No minimum
- No maximum
- Domain: x > −3 or (−3, ∞)
- Range: All Real Numbers $\mathbb{R}$ or (−∞, ∞)
- Quadrant I, II & III
- Parent function is $f(x) = \ln x$

FOD #95 $\qquad f(x) = 5e^x$

- Exponential Growth Function
- Base is "e"
- Continuous
- Increasing
- Vertical stretch by a factor of 5
- y-intercept is (0, 5)
- No x-intercept
- Horizontal Asymptote y = 0
- Domain: All Real Numbers $\mathbb{R}$ or (–∞, ∞)
- Range: y > 0 or (0, ∞)
- Quadrants I & II
- Parent function is $f(x) = e^x$

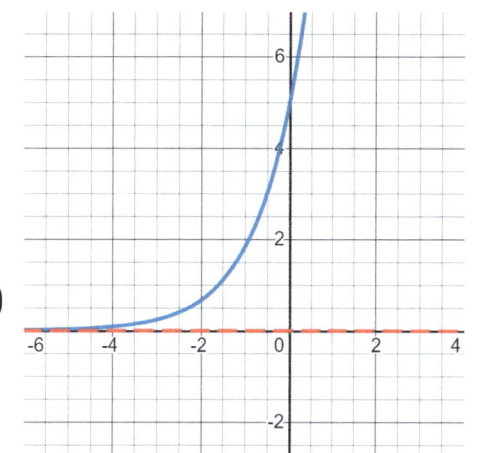

© Debra Richardson 2022

FOD #96 $\qquad f(x) = -4e^x$

- Exponential Growth Function
- Base is "e"
- Continuous
- Reflects over x-axis
- Decreasing
- Vertical stretch by a factor of 4
- y-intercept is (0, –4)
- No x-intercept
- Horizontal Asymptote y = 0
- Domain: All Real Numbers $\mathbb{R}$ or (–∞, ∞)
- Range: y < 0 or (–∞, 0)
- Quadrants III & IV
- Parent function is $f(x) = e^x$

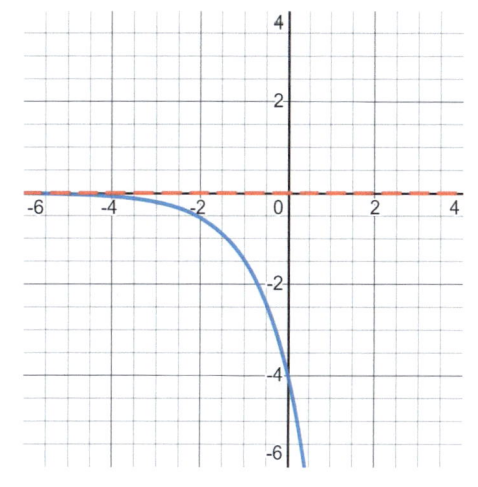

© Debra Richardson 2022

FOD #97

$$f(x) = 6e^{-x}$$

- Exponential Growth Function
- Base is "e"
- Continuous
- Reflects over y-axis
- Decreasing
- Vertical stretch by a factor of 6
- y-intercept is (0, 6)
- No x-intercept
- Horizontal Asymptote y = 0
- Domain: All Real Numbers $\mathbb{R}$ or (–∞, ∞)
- Range: y > 0 or (0, ∞)
- Quadrants I & II
- Parent function is $f(x) = e^x$

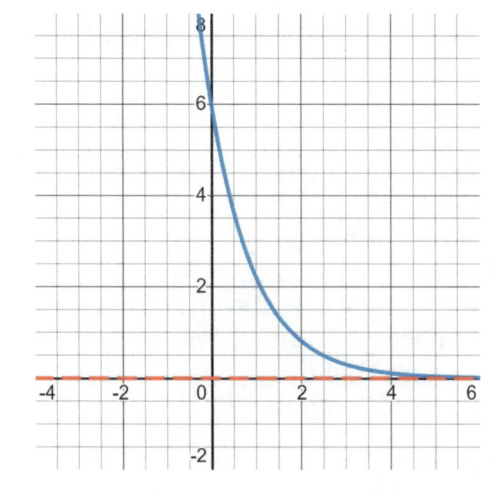

© Debra Richardson 2022

FOD #98

$$f(x) = \sqrt[3]{x}$$

- Cube Root Parent Function
- Radical Function
- Relation
- Continuous
- Nonlinear
- Increasing
- No maximum, no minimum
- y-intercept is (0, 0)
- x-intercept is (0, 0)
- Domain: All Real Numbers $\mathbb{R}$ or (–∞, ∞)
- Range: All Real Numbers $\mathbb{R}$ or (–∞, ∞)
- Quadrants I & III
- End Behavior: As x→ –∞, f(x)→ –∞
- End Behavior: As x→ +∞, f(x)→ +∞

© Debra Richardson 2022

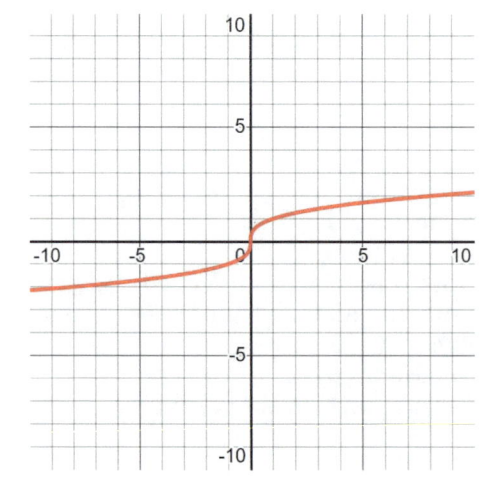

FOD #99 $f(x) = \sqrt[3]{x} + 2$

- Cube Root Function
- Continuous
- Increasing
- No maximum, no minimum
- Vertical shift up 2
- y-intercept is (0, 2)
- x-intercept is (–8, 0)
- Domain: All Real Numbers $\mathbb{R}$ or (–∞, ∞)
- Range: All Real Numbers $\mathbb{R}$ or (–∞, ∞)
- Quadrants I, II & III
- End Behavior: As x→ –∞, f(x)→ –∞
- End Behavior: As x→ +∞, f(x)→ +∞
- Inverse function $f^{-1}(x) = (x-2)^3$
- Parent function is $f(x) = \sqrt[3]{x}$

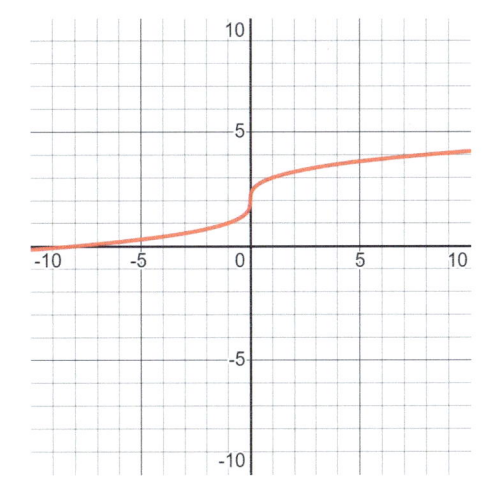

FOD #100 Longhorn $f(x) = \left|\sqrt[3]{x}\right|$ $g(x) = -\left|\dfrac{2}{x}\right|$

- Each piece is a function
- Each piece is continuous
- Longhorn shape would <u>NOT</u> be a function
- Symmetry over y-axis
- $f(x)$ in quadrants I & II
- $g(x)$ in quadrants III & IV
- $f(x)$ touches the origin
- x and y-intercept is (0, 0) for $f(x)$
- Asymptotes at x = 0 and y = 0 for $g(x)$
- Domain of $f(x)$: $\mathbb{R}$ or(–∞, ∞)
- Range of $f(x)$: y ≥ 0 or [0, ∞]
- Domain of $g(x)$: (–∞, 0) U (0, ∞)
- Range of $g(x)$: (–∞, 0)

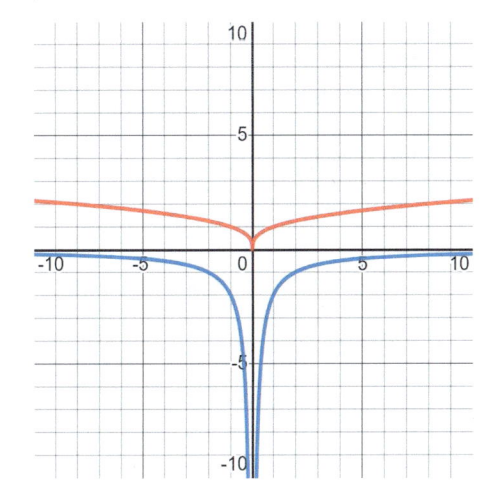

FOD #101

$$f(x) = \sqrt[3]{x-8} + 2$$

- Cube Root Function
- Continuous
- Increasing
- No maximum, no minimum
- Horizontal shift right 8
- Vertical shift up 2
- y-intercept is (0, 0)
- x-intercept is (0, 0)
- Domain: All Real Numbers $\mathbb{R}$ or (–∞, ∞)
- Range: All Real Numbers $\mathbb{R}$ or (–∞, ∞)
- Quadrants I & III
- End Behavior: As x→ –∞, f(x)→ –∞
- End Behavior: As x→ +∞, f(x)→ +∞
- Parent function is $f(x) = \sqrt[3]{x}$

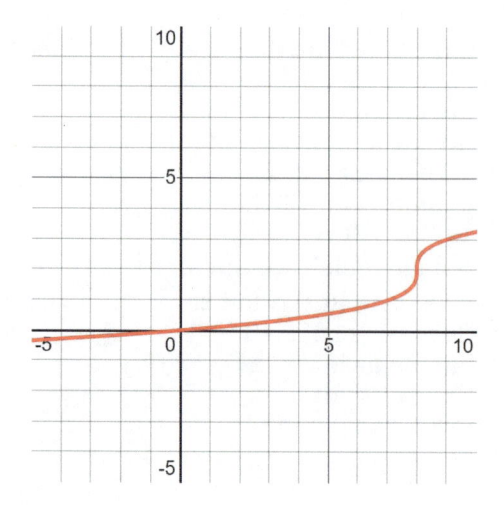

© Debra Richardson 2022

FOD #102

$$f(x) = -\sqrt[3]{x+2} - 1$$

- Cube Root Function
- Reflects over x-axis
- Decreasing
- Horizontal shift left 2
- Vertical shift down 1
- y-intercept is (0, –2.26)
- x-intercept is (–3, 0)
- Domain: All Real Numbers $\mathbb{R}$ or (–∞, ∞)
- Range: All Real Numbers $\mathbb{R}$ or (–∞, ∞)
- Quadrants II, III & IV
- End Behavior: As x→ –∞, f(x)→ +∞
- End Behavior: As x→ +∞, f(x)→ –∞
- Parent function is $f(x) = \sqrt[3]{x}$

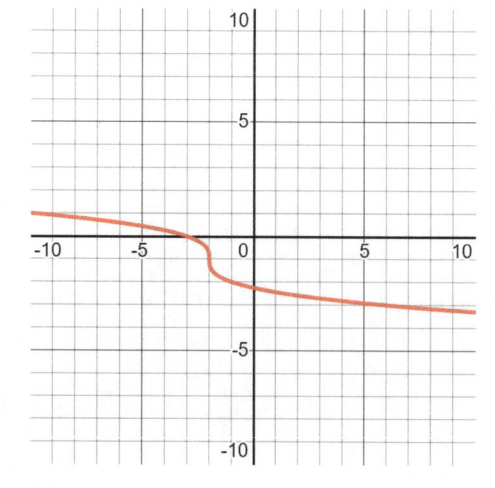

© Debra Richardson 2022

FOD #103

$$f(x) = -2\sqrt[3]{x+3}$$

- Cube Root Function
- Reflects over x-axis
- Decreasing
- Vertical stretch by a factor of 2
- Horizontal shift left 3
- y-intercept is (0, –2.88)
- x-intercept is (–3, 0)
- Domain: All Real Numbers ℝ or (–∞, ∞)
- Range: All Real Numbers ℝ or (–∞, ∞)
- Quadrants II, III & IV
- End Behavior: As x→ –∞, f(x)→ +∞
- End Behavior: As x→ +∞, f(x)→ –∞
- Parent function is $f(x) = \sqrt[3]{x}$

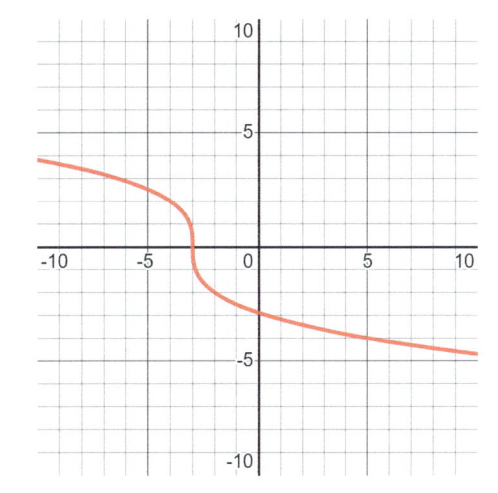

© Debra Richardson 2022

FOD #104

$$f(x) = \frac{1}{2}\sqrt[3]{(x+4)} - 1$$

- Cube Root Function
- Continuous
- Increasing
- Vertical compression by a factor of ½
- Horizontal shift left 4
- Vertical shift down 1
- y-intercept is (0, –0.2)
- x-intercept is (4, 0)
- Domain: All Real Numbers ℝ or (–∞, ∞)
- Range: All Real Numbers ℝ or (–∞, ∞)
- Quadrants I, III & IV
- End Behavior: As x→ –∞, f(x)→ –∞
- End Behavior: As x→ +∞, f(x)→ +∞
- Parent function is $f(x) = \sqrt[3]{x}$

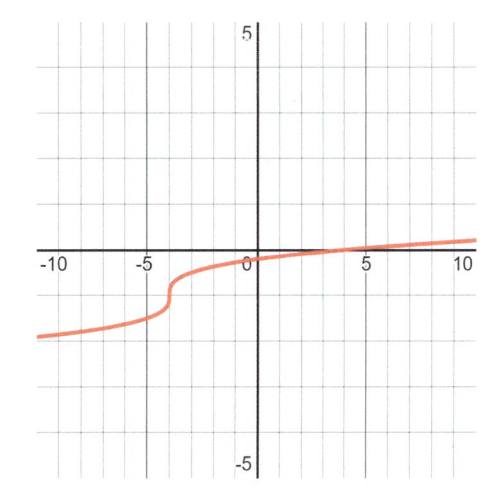

© Debra Richardson 2022

FOD #105

$$f(x) = \cos(x)$$

- Cosine Function
- Parent Function
- Trigonometric Function
- Periodic Function, Oscillating
- Increases and Decreases
- Continuous
- Amplitude is 1
- Period is 2π
- Midline is y = 0
- y-intercept is (0, 1)
- x-intercepts are at $(\ldots -\frac{3\pi}{2}, -\frac{\pi}{2}, \frac{\pi}{2}, \frac{3\pi}{2} \ldots)$
- Domain: All Real Numbers ℝ or (–∞, ∞)
- Range: –1 ≤ y ≤ 1 or [–1, 1]
- All Quadrants

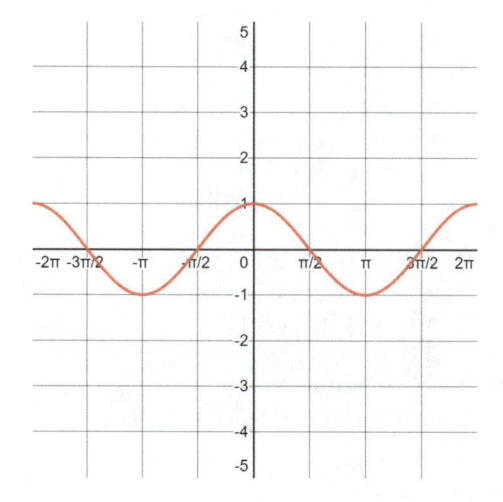

© Debra Richardson 2022

FOD #106

$$f(x) = -\cos(x)$$

- Cosine Function
- Reflects over x-axis
- Increases and Decreases
- Continuous
- Amplitude is 1
- Period is 2π
- Midline is y = 0
- y-intercept is (0, –1)
- x-intercepts are at $(\ldots -\frac{3\pi}{2}, -\frac{\pi}{2}, \frac{\pi}{2}, \frac{3\pi}{2} \ldots)$
- Domain: All Real Numbers ℝ or (–∞, ∞)
- Range: –1 ≤ y ≤ 1 or [–1, 1]
- All Quadrants
- Parent Function is $f(x) = \cos(x)$

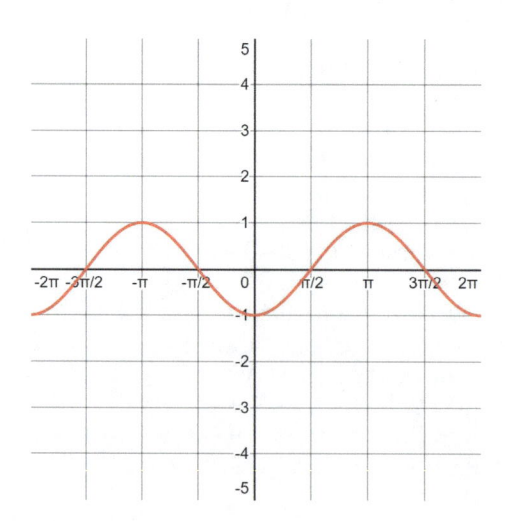

© Debra Richardson 2022

FOD #107　　　$f(x) = \cos(x) - 1$

- Cosine Function
- Vertical shift down 1
- Decreases and Increases
- Continuous
- Amplitude is 1
- Period is 2π
- Midline is y = −1
- y-intercept is (0, 0)
- x-intercepts are at (… −2π, 0, 2π …)
- Domain: All Real Numbers ℝ or (−∞, ∞)
- Range: − 2 ≤ y ≤ 0 or [− 2, 0]
- Quadrants III & IV
- Parent Function $f(x) = \cos(x)$

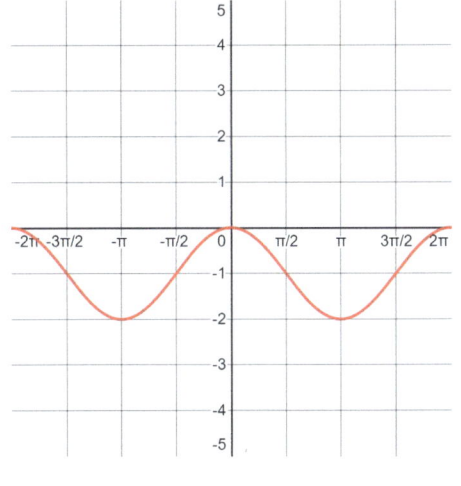

FOD #108　　　$f(x) = 3\cos(x)$

- Cosine Function
- Increases and Decreases
- Continuous
- Vertical stretch by a factor of 3
- Amplitude is 3
- Period is 2π
- Midline is y = 0
- y-intercept is (0, 3)
- x-intercepts are at $(\ldots -\frac{3\pi}{2}, -\frac{\pi}{2}, \frac{\pi}{2}, \frac{3\pi}{2} \ldots)$
- Domain: All Real Numbers ℝ or (−∞, ∞)
- Range: −3 ≤ y ≤ 3 or [−3, 3]
- All Quadrants
- Parent Function is $f(x) = \cos(x)$

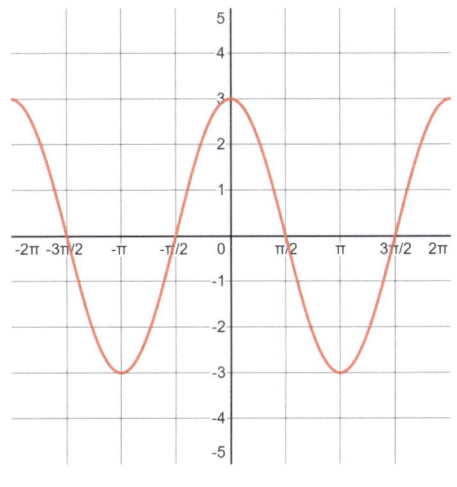

FOD #109

$$f(x) = -2\cos(x)$$

- Cosine Function
- Reflects over x-axis
- Increases and Decreases
- Continuous
- Vertical stretch by a factor of 2
- Amplitude is 2
- Period is 2π
- Midline is y = 0
- y-intercept is (0, −2)
- x-intercepts are at $(\ldots -\frac{3\pi}{2}, -\frac{\pi}{2}, \frac{\pi}{2}, \frac{3\pi}{2} \ldots)$
- Domain: All Real Numbers ℝ or (−∞, ∞)
- Range: −2 ≤ y ≤ 2 or [−2, 2]
- All Quadrants
- Parent Function $f(x) = \cos(x)$

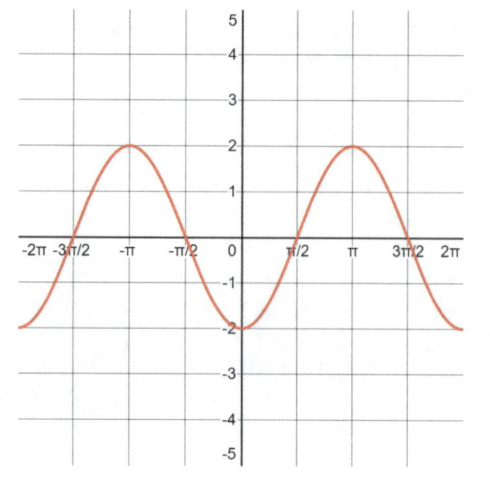

© Debra Richardson 2022

FOD #110

$$f(x) = \sin(x)$$

- Sine Function
- Parent Function
- Trigonometric Function
- Periodic Function, Oscillating
- Increases and Decreases
- Continuous
- Amplitude is 1
- Period is 2π
- Midline is y = 0
- y-intercept is (0, 0)
- x-intercepts are at (… −2π, −π, 0, π, 2π …)
- Domain: All Real Numbers ℝ or (−∞, ∞)
- Range: −1 ≤ y ≤ 1 or [−1, 1]
- All Quadrants

© Debra Richardson 2022

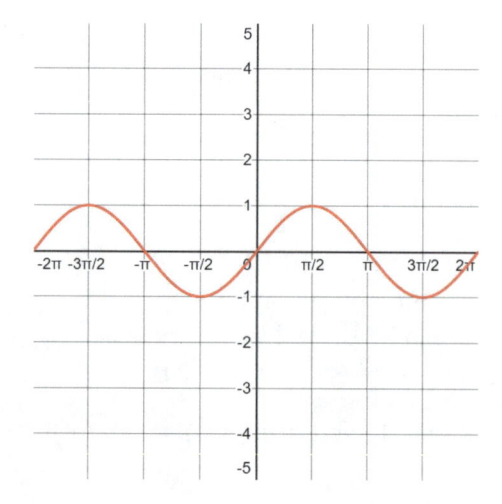

FOD #111

$$f(x) = \sin(x) + 2$$

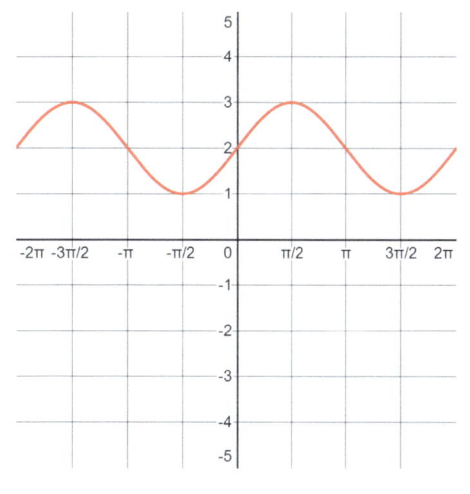

- Sine Function
- Vertical shift up 2
- Increases and Decreases
- Continuous
- Amplitude is 1
- Period is 2π
- Midline is y = 2
- y-intercept is (0, 2)
- No x-intercepts
- Domain: All Real Numbers ℝ or (−∞, ∞)
- Range: 1 ≤ y ≤ 3 or [1, 3]
- Quadrants I & II
- Parent Function is $f(x) = \sin x$

© Debra Richardson 2022

FOD #112

$$f(x) = \sin(x + \pi)$$

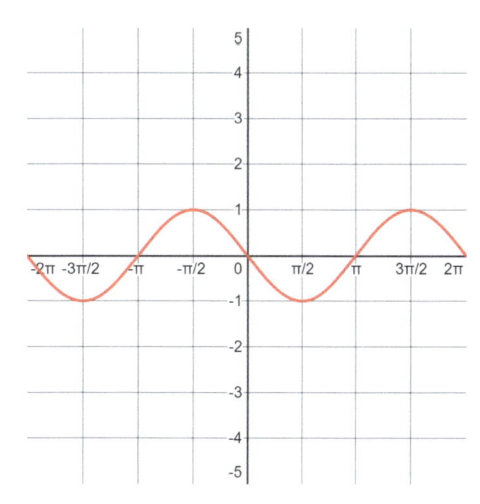

- Sine Function
- Horizontal shift left π
- Increases and Decreases
- Continuous
- Amplitude is 1
- Period is 2π
- Midline is y = 0
- y-intercept is (0, 0)
- x-intercepts are at (… −2π, −π, 0, π, 2π …)
- Domain: All Real Numbers ℝ or (−∞, ∞)
- Range: −1 ≤ y ≤ 1 or [−1, 1]
- All Quadrants
- Parent Function is $f(x) = \sin x$

© Debra Richardson 2022

FOD #113

$$f(x) = \sin(x - \pi) - 2$$

- Sine Function
- Increases and Decreases
- Continuous
- Amplitude is 1
- Period is 2π
- Horizontal shift right π
- Vertical shift down 2
- Midline is y = –2
- y-intercept is (0, –2)
- No x-intercepts
- Domain: All Real Numbers $\mathbb{R}$ or (–∞, ∞)
- Range: –3 ≤ y ≤ –1 or [–3, –1]
- Quadrants III & IV
- Parent Function is $f(x) = \sin x$

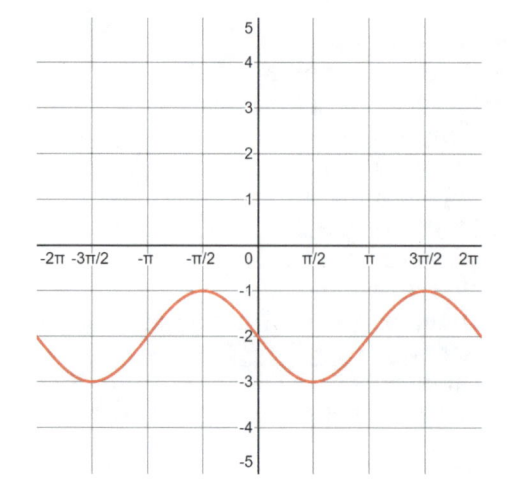

© Debra Richardson 2022

FOD #114

$$f(x) = 3\sin(2x)$$

- Sine Function
- Increases and Decreases
- Continuous
- Vertical stretch by a factor of 3
- Amplitude is 3
- Horizontal compression by a factor of ½
- Period is π
- Midline is y = 0
- y-intercept is (0, 0)
- x-intercepts are at (… $-\pi, -\pi/2, 0, \pi/2, \pi$ …)
- Domain: All Real Numbers $\mathbb{R}$ or (–∞, ∞)
- Range: –3 ≤ y ≤ 3 or [–3, 3]
- All Quadrants
- Parent Function is $f(x) = \sin x$

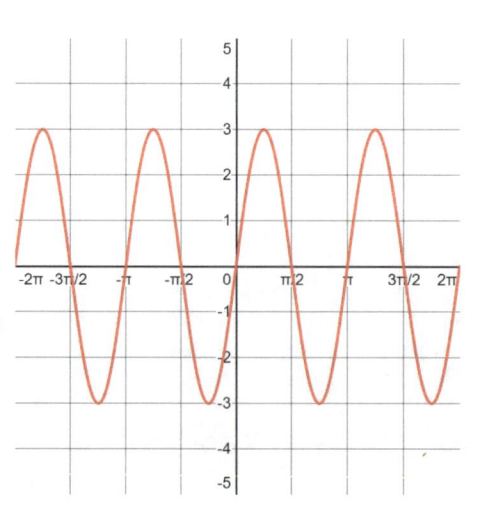

© Debra Richardson 2022

FOD #115

$f(x) = 4\sin(x) \quad g(x) = -4\sin(x)$

- Sine Functions
- Increasing and Decreasing
- Continuous
- Vertical stretch by a factor of 4
- Amplitude is 4
- Period is 2π
- Midline is y = 0
- y-intercept is (0, 0)
- x-intercepts are at $(\ldots -2\pi, -\pi, 0, \pi, 2\pi \ldots)$
- Intersections at $(\ldots -2\pi, -\pi, 0, \pi, 2\pi \ldots)$
- Domain: All Real Numbers $\mathbb{R}$ or $(-\infty, \infty)$
- Range: $-4 \leq y \leq 4$ or $[-4, 4]$
- All Quadrants

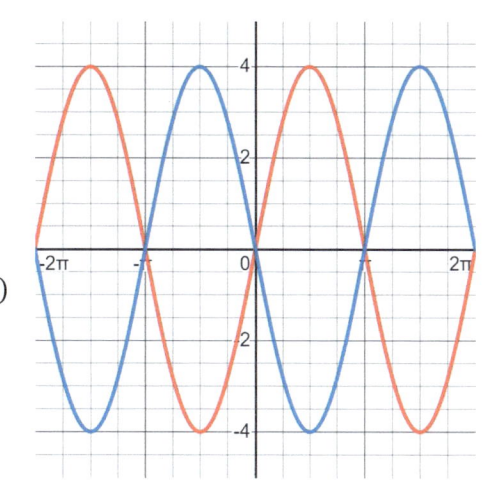

© Debra Richardson 2022

FOD #116

$f(x) = \tan(x)$

- Tangent Function
- Parent Function
- Trigonometric Function
- Periodic Function
- Increasing
- Continuous over the domain
- Period is π
- Vertical Asymptotes at x = $\ldots -\frac{3\pi}{2}, -\frac{\pi}{2}, \frac{\pi}{2}, \frac{3\pi}{2} \ldots$
- y-intercept is (0, 0)
- x-intercepts are at $(\ldots -2\pi, -\pi, 0, \pi, 2\pi \ldots)$
- Domain: $\{x \in \mathbb{R} \mid x \neq \ldots -\frac{3\pi}{2}, -\frac{\pi}{2}, \frac{\pi}{2}, \frac{3\pi}{2} \ldots\}$
- Range: $\mathbb{R}$
- All Quadrants

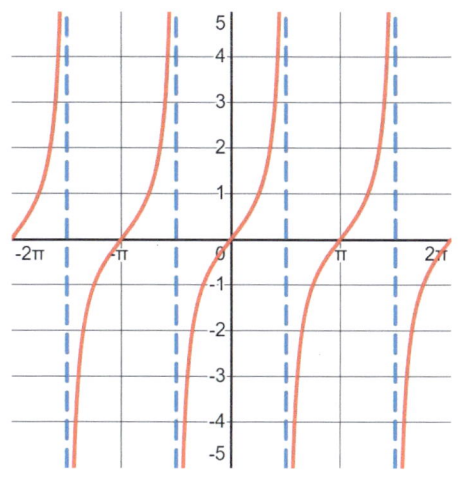

© Debra Richardson 2022

FOD #117

$f(x) = -\tan(x)$

- Tangent Function
- Trigonometric Function
- Reflects over x-axis
- Decreasing
- Continuous over the domain
- Period is π
- Vertical Asymptotes at x = ... $-\frac{3\pi}{2}, -\frac{\pi}{2}, \frac{\pi}{2}, \frac{3\pi}{2}$ ...
- y-intercept is (0, 0)
- x-intercepts are at (...−2π, −π, 0, π, 2π ...)
- Domain: {x ϵ ℝ | x ≠ ... $-\frac{3\pi}{2}, -\frac{\pi}{2}, \frac{\pi}{2}, \frac{3\pi}{2}$ ...}
- Range: ℝ
- All Quadrants
- Parent function is $f(x) = \tan(x)$

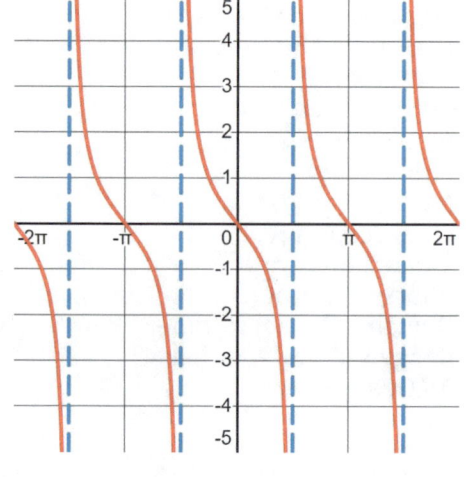

© Debra Richardson 2022

FOD #118

$f(x) = \tan(x/2)$

- Tangent Function
- Horizontal stretch by a factor of 2
- Increasing
- Continuous over the domain
- Period is 2π
- Vertical Asymptotes at x =... −3π, −π, π, 3π ...
- y-intercept is (0, 0)
- x-intercepts are at (... −4π, −2π, 0, 2π, 4π ...)
- Domain: {x ϵ ℝ | x ≠ ... −3π, −π, π, 3π ... }
- Range: ℝ
- All Quadrants
- Parent Function is $f(x) = \tan(x)$

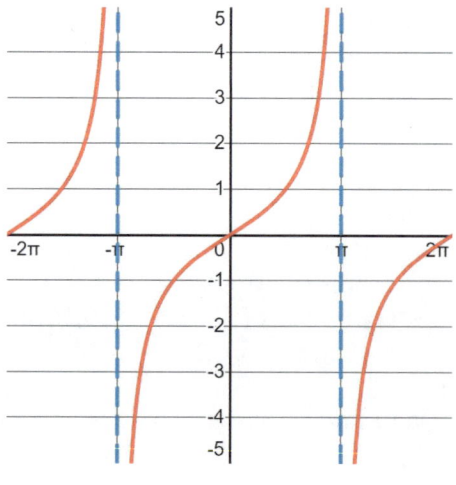

© Debra Richardson 2022

FOD #119

$f(x) = \tan(2x)$

- Tangent Function
- Trigonometric Function
- Horizontal compression by a factor of ½
- Increasing
- Continuous over the domain
- Period is $\frac{\pi}{2}$
- Vertical Asymptotes at x = ... $-\frac{3\pi}{4}, -\frac{\pi}{4}, \frac{\pi}{4}, \frac{3\pi}{4}$ ...
- y-intercept is (0, 0)
- x-intercepts are at (... $-\frac{3\pi}{2}, -\frac{\pi}{2}, 0, \frac{\pi}{2}, \frac{3\pi}{2}$ ...)
- Domain: {x ϵ ℝ | x ≠ ... $-\frac{3\pi}{4}, -\frac{\pi}{4}, \frac{\pi}{4}, \frac{3\pi}{4}$ ...}
- Range: ℝ
- All Quadrants
- Parent Function is $f(x) = \tan(x)$

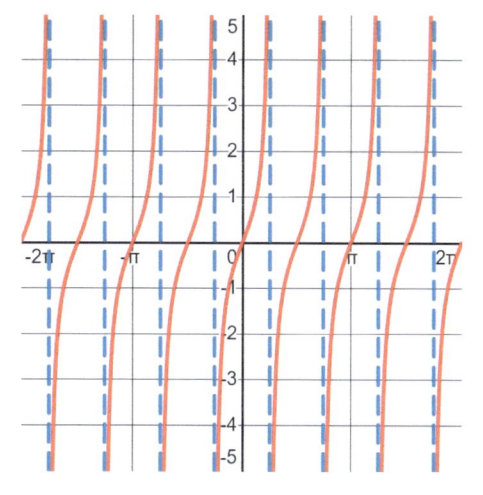

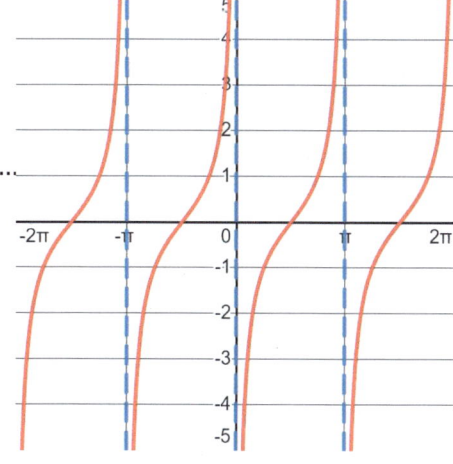

© Debra Richardson 2022

FOD #120

$f(x) = \tan(x - \pi/2)$

- Tangent Function
- Horizontal shift right π/2
- Increasing
- Continuous over the domain
- Period is π
- Vertical Asymptotes at x = ... −2π, −π, 0, π, 2π ...
- No y-intercept
- x-intercepts are at (... $-\frac{3\pi}{2}, -\frac{\pi}{2}, \frac{\pi}{2}, \frac{3\pi}{2}$, ...)
- Domain: {x ϵ ℝ | x ≠ ... −3π, −π, 0, π, 3π ... }
- Range: ℝ
- All Quadrants
- Parent Function is $f(x) = \tan(x)$

© Debra Richardson 2022

FOD #121
for Valentine's Day

$y = \sqrt{25 - (x+5)^2}$
$y = \sqrt{25 - (x-5)^2}$
$y = x - 10, \ 0 \geq x \geq 10$
$y = -x - 10, -10 \geq x \geq 0$

- Each piece is a function
- Heart shape: 2 semi-circles & 2 lines
- Heart shape would NOT be a function
- Relation
- Touches the origin
- Symmetry over y-axis
- Maximum is 5
- Minimum is –10
- y-intercepts (0, 0), (0, –10)
- x-intercepts (–10, 0), (0, 0), (10, 0)
- Domain: [–10, 10]
- Range: [–10, 5]
- All Quadrants

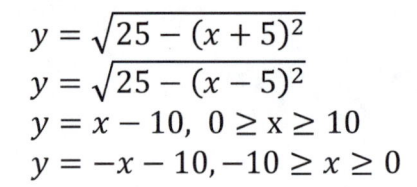

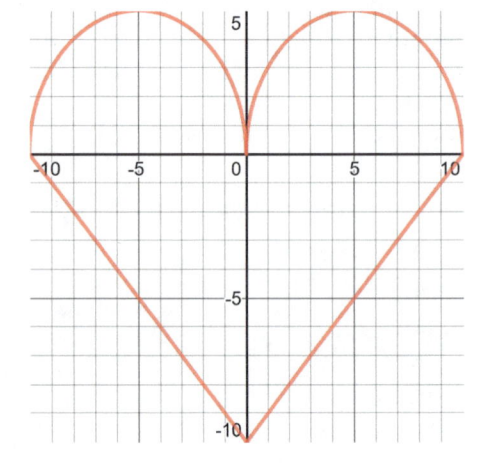

© Debra Richardson 2022

FOD #122
for Pi Week

$f(r) = 2\pi r, r > 0$

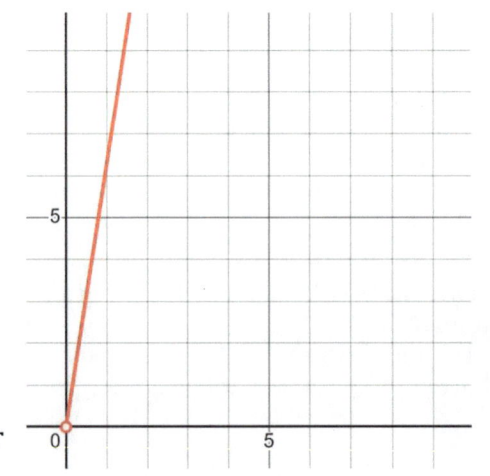

- Linear Function
- Relation
- Continuous
- Does not include the origin
- Increasing
- Slope is 2π
- No intercepts
- Domain: r > 0 or (0, ∞)
- Range: $f(r) > 0$ or (0, ∞)
- Quadrant I
- $f(r)$ is circumference of circle with radius r

© Debra Richardson 2022

FOD #123 for Pi Week

$$f(r) = \pi r^2, r > 0$$

- Quadratic Function
- Relation
- Continuous
- Nonlinear
- Does not include the origin
- Increasing
- No intercepts
- Domain: r > 0 or $(0, \infty)$
- Range: $f(r) > 0$ or $(0, \infty)$
- Quadrant I
- $f(r)$ is area of a circle with radius r

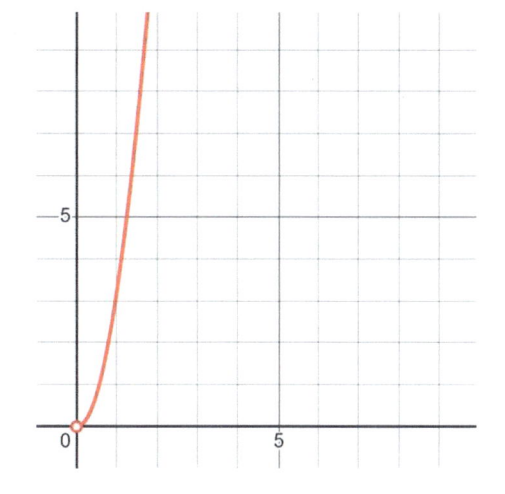

© Debra Richardson 2022

FOD #124 for Pi Week

$$f(r) = \frac{4}{3}\pi r^3, r > 0$$

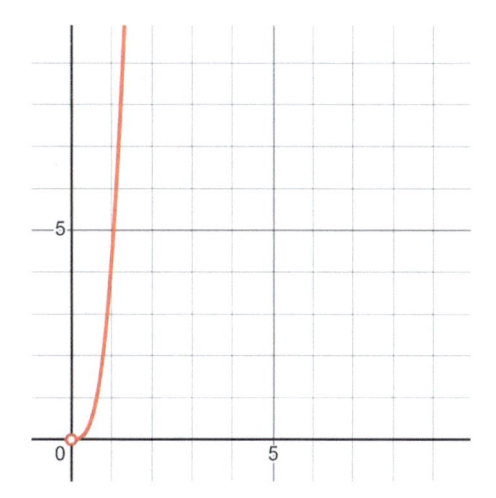

- Cubic Function
- Relation
- Continuous
- Nonlinear
- Does not include the origin
- Increasing
- No intercepts
- Domain: r > 0 or $(0, \infty)$
- Range: $f(r) > 0$ or $(0, \infty)$
- Quadrant I
- $f(r)$ is volume of sphere with radius r

© Debra Richardson 2022

FOD #125
Extra Linear System

$y = \frac{3}{2}x + 6$

$y = 2 - \frac{x}{2}$

- Two Linear Functions
- System of Linear Equations
- Relations
- Continuous
- Lines intersect at (–2, 3)
- Solution to the system is (–2, 3)
- Increasing and Decreasing Lines
- Slopes are 3/2 and –1/2
- y-intercepts are (0, 6) and (0, 2)
- x-intercepts are (–4, 0) and (4, 0)
- Each Domain: All Real Numbers $\mathbb{R}$ or (–∞, ∞)
- Each Range: All Real Numbers $\mathbb{R}$ or (–∞, ∞)

© Debra Richardson 2022

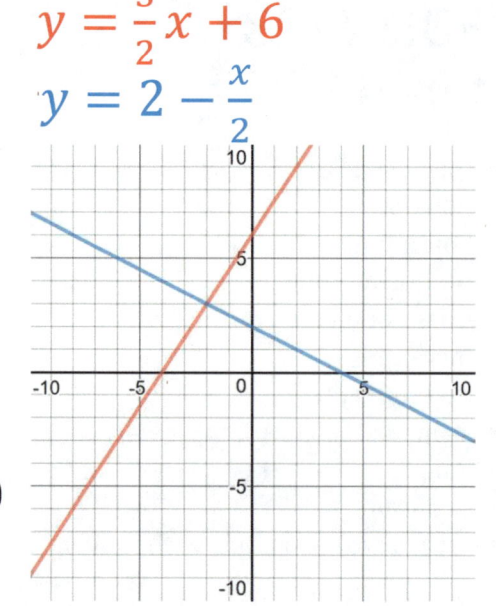

FOD #126
Extra Quadratic

$f(x) = 2x^2 + 3x - 5$

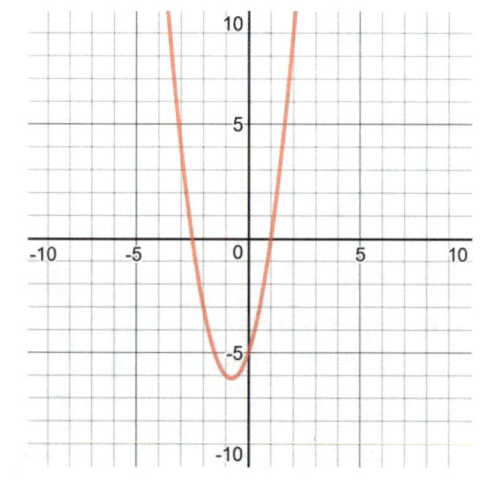

- Quadratic Function
- Factored Form is y = (2x+5)(x – 1)
- Relation, Continuous
- Parabola opens upward
- y-intercept is at (0, –5)
- x-intercepts are (–2.5, 0) and (1, 0)
- Axis of Symmetry is x = –0.75
- Vertex is (–0.75, –6.125)
- Minimum point is (–0.75, –6.125)
- Domain: All Real Numbers $\mathbb{R}$ or (–∞, ∞)
- Range: y ≥ –6.125 or [–6.125, ∞)
- All Quadrants
- Decreasing Interval (–∞, –0.75)
- Increasing Interval (–0.75, ∞)
- Parent Function is $f(x) = x^2$

© Debra Richardson 2022

FOD #127
Extra Cubic

$$f(x) = x^3 - 1$$

- Cubic Function
- Relation
- Continuous
- Nonlinear
- Vertical shift down 1
- Increasing
- No maximum, no minimum
- y-intercept is (0, –1)
- x-intercept is (1, 0)
- Domain: All Real Numbers $\mathbb{R}$ or (–∞, ∞)
- Range: All Real Numbers $\mathbb{R}$ or (–∞, ∞)
- Quadrants I, III & IV
- End Behavior: As x→ –∞, f(x)→ –∞
- End Behavior: As x→ +∞, f(x)→ +∞
- Parent Function is $f(x) = x^3$

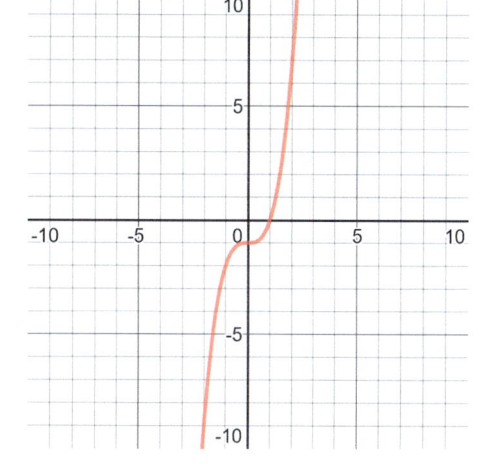

© Debra Richardson 2022

FOD #128
Extra Quartic

$$f(x) = x(x - 1)(x + 2)(x - 3)$$

- Quartic Function
- Degree 4 Polynomial
- Opens upward
- Has 3 turning points
- Relative Maximum is apx. (0.5, 1.6)
- Relative Minima are apx. (–1.3, –9) & (2.3, –9)
- y-intercept is (0, 0)
- x-intercepts are (–2, 0), (0, 0), (1, 0), (3, 0)
- Domain: All Real Numbers $\mathbb{R}$ or (–∞, ∞)
- Range: y ≥ – 9 or [–9, ∞)
- All Quadrants
- End Behavior: As x → –∞, f(x) → +∞
- End Behavior: As x → +∞, f(x) → +∞
- Parent Function is $f(x) = x^4$

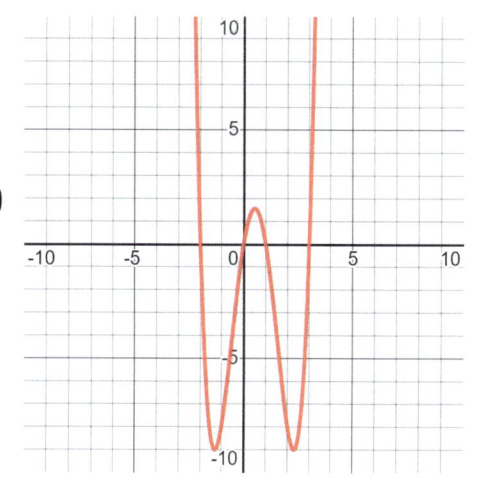

© Debra Richardson 2022

FOD #129
Extra Exponential

$$f(x) = -8\left(\frac{1}{2}\right)^x$$

- Exponential Decay Function
- Base is ½
- Reflects over x-axis
- Vertical stretch by a factor of 8
- Increasing
- y-intercept is (0, –8)
- No x-intercepts
- Horizontal Asymptote y = 0
- Never touches x-axis
- Domain: All Real Numbers ℝ or (–∞, ∞)
- Range: y < 0 or (–∞, 0)
- Quadrants III & IV
- Parent Function is $f(x) = (b)^x$

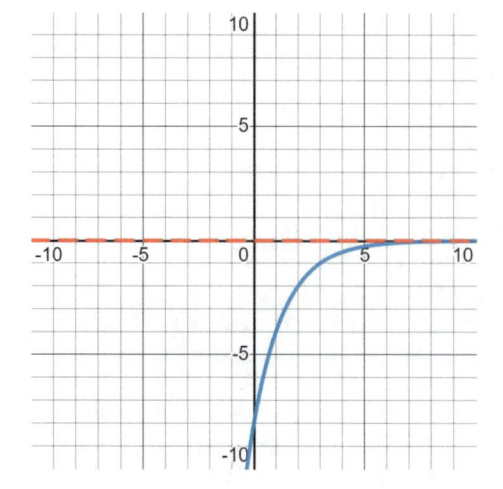

© Debra Richardson 2022

FOD #130
Extra Exponential

$$f(x) = 8\left(\frac{1}{2}\right)^{-x}$$

- Exponential Decay Function
- Base is ½
- Reflects over y-axis
- Vertical stretch by a factor of 8
- Increasing
- y-intercept is (0, 8)
- No x-intercepts
- Horizontal Asymptote y = 0
- Never touches x-axis
- Domain: All Real Numbers ℝ or (–∞, ∞)
- Range: y > 0 or (0, ∞)
- Quadrants I & II
- Parent Function is $f(x) = (b)^x$

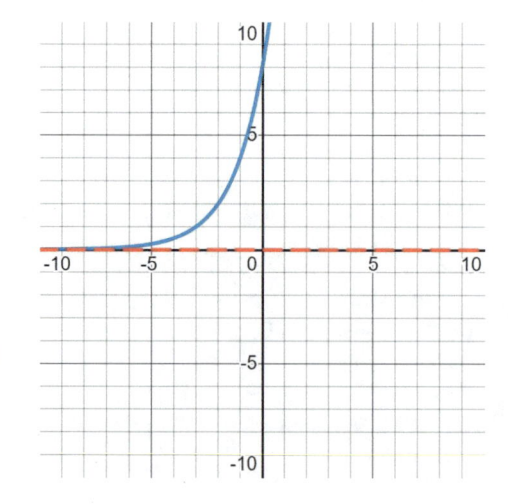

© Debra Richardson 2022

VOCABULARY   #FOD

Function number when introduced

Absolute Value	#	29
Absolute Value Function	#	29
Amplitude	#	105
Area	#	123
Asymptote	#	58
Axes	#	3
Axis of Symmetry	#	17
Base	#	81
Circle	#	122
Circumference	#	122
Compression	#	104
Constant	#	8
Continuous	#	1
Cosine Function	#	105
Cube Root	#	98
Cube Root Function	#	98
Cubic	#	45
Cubic Function	#	45
Decreasing	#	2
Decreasing Interval	#	17
Degree	#	45
Domain	#	1
Double Root	#	20
End Behavior	#	45
Euler's Number	#	89
Even Function	#	17
Exponent	#	45
Exponential "e" Function	#	89
Exponential Decay	#	71
Exponential Function	#	70
Exponential Growth	#	70
Factored Form	#	27
Function	#	1
Hole	#	66
Horizontal	#	8
Horizontal Asymptote	#	58
Horizontal Compression	#	114
Horizontal Shift	#	20
Horizontal Stretch	#	118
Increasing	#	1
Increasing Interval	#	17

Infinity	#	1
Initial Amount	#	73
Intersect	#	13
Interval Notation	#	58
Inverse	#	80
Inverse Function	#	80
Irrational Number	#	122
Leading Coefficient	#	56
Linear	#	1
Linear Equation	#	13
Linear Function	#	1
Logarithmic	#	81
Logarithmic Function	#	81
Maximum	#	18
Midline	#	105
Minimum	#	17
Narrow	#	23
Natural Log	#	90
Natural Log Function	#	90
Negative	#	74
Nonlinear	#	17
Odd Function	#	45
Opposite Reciprocal	#	13
Origin	#	1
Oscillating	#	105
Parabola	#	17
Parallel	#	15
Parent Function	#	1
Period	#	105
Periodic Function	#	105
Perpendicular	#	13
Pi	#	122
Polynomial	#	45
Quadrant I	#	1
Quadrant II	#	2
Quadrant III	#	1
Quadrant IV	#	2
Quadrants	#	1
Quadratic	#	17
Quadratic Function	#	17
Quartic Function	#	53
Radical	#	37
Radical Function	#	37
Radius	#	122

Range	#	1
Rate of Decay	#	71
Rate of Growth	#	70
Ratio	#	58
Rational	#	58
Rational Function	#	58
Real Number	#	1
Reflection	#	18
Relation	#	1
Relative Maxima	#	56
Relative Maximum	#	51
Relative Minima	#	55
Relative Minimum	#	51
Removable Discontinuity	#	66
Root	#	20
Set Notation	#	58
Simplified Equation	#	66
Sine Function	#	110
Slope	#	1
Slope-Intercept Form	#	3
Solution to System	#	13
Sphere	#	124
Square Root	#	37
Square Root Function	#	37
Standard Form	#	9
Stretch	#	44
Symmetry	#	17
Systems of Equations	#	13
Tangent Function	#	116
Trigonometric Function	#	105
Turning Points	#	50
Union	#	58
Vertex	#	17
Vertical	#	8
Vertical Asymptote	#	58
Vertical Compression	#	24
Vertical Shift	#	19
Vertical Stretch	#	23
Volume	#	124
Wide	#	24
x-axis	#	3
x-intercept	#	1
y-axis	#	3
y-intercept	#	1

Made in the USA
Columbia, SC
28 May 2025